PF

THIS book began its life as the Lees Knowles lectures, delivered in Cambridge in January and February 1995. My first debt therefore is to the Master and Fellows of Trinity College for their invitation, and my second is to those who ventured out to the Mill Lane lecture rooms to listen and to question. Some of the points raised then have found reflection in this much expanded version of what I said.

When I was first asked to give these lectures I was still a resident Fellow of Corpus Christi College. The fact that in the interim I migrated to Glasgow gave me and my wife—on the occasions when her other heavy obligations permitted—the opportunity to savour the splendid delights of Trinity's hospitality. In particular I must thank Dr Boyd Hilton, who made all the arrangements for the lectures, and Professor Robert Neild, who presided over them and us with the solicitude of long-standing friendship. Professor Derek Beales, in addition to being one of my most persistent auditors, organized a dinner party in Sidney Sussex, and the return to Cambridge allowed me some time to enjoy the familiar and enfolding embrace of the Life Fellowship to which Corpus elected me on my departure in 1992.

Giving (and, I have to confess, frantically writing) a series of lectures at the other end of the country in the course of term could not be done without some neglect of my duties in Glasgow. My colleagues did not complain, despite dummy runs for quality assurance and research assessment. Dr Simon Dixon took over my Ordinary class teaching. Dr Lionel Glassey and Dr Simon Ball made bibliographical suggestions in their respective areas of expertise. My prime Glasgow debt, however, must be to Patricia Ferguson, who typed the manuscript of this book with speed, with an accuracy that my legibility does not warrant, and with an enthusiasm for the outcome that was most encouraging—and some of it beyond her formal retirement date.

Sir Lees Knowles, the benefactor who endowed the lectures that bear his name, was educated at Rugby. When Lord Hankey gave the Lees Knowles lectures in 1945, under the title *Government Control in War*, he made much of the fact that he was the first Rugbeian to do so. He was wrong. C. R. M. F. Cruttwell, the historian of the First World War, lectured on British strategy in that war in 1936. Since then at least one other, Donald Cameron Watt, has done so, in a series which I was privileged to hear as a new research student and which was published as *Too Serious a Business: European Armed Forces and the Approach to the Second World War*.

I too was educated at Rugby. It is less fashionable to make such connections now than it was in Hankey's day, and certainly I am conscious of my limitations as I find myself in such distinguished company. Nonetheless I am struck by the fact that all four of us have chosen topics that constitute cognate themes. Explanations that seek a common bond in the Midlands are elusive. Rugby has produced fewer great politicians and even fewer great soldiers than its reputation might suggest. Only one bulks large in these pages—Sir Henry Pownall—but he was a soldier who most emphatically confirms the central thesis of this book.

The title of the book is more descriptive than arresting. Certainly it is not original. In 1955 Gordon Craig wrote a path-breaking study that became both a standard text and an exemplar for subsequent research in the same genre, *The Politics of the Prussian Army*. In the 1960s and 1970s, as military history staged its claim to academic respectability, it did so not on the back of the operational history beloved by official historians but on studies of war and society. Armies were interesting in so far as their composition did or did not reflect the civil societies from which they were drawn; their high commands and general staffs were important not for their direction of armed forces in wartime but for their propensity to intervene in the machinery of government at any time.

These were fruitful lines of enquiry. Military historians learned from sociologists and political scientists; political scientists and sociologists plundered the past for the evidence to support their theories. The vocabulary that emerged was symbolic of this cross-fertilization. 'Bonapartism' was no longer just a description of a French political movement, of those wedded to the restoration to France of Napoleon and his heirs. It was also used more generally, to describe the ability to overcome latent domestic divisions

THE POLITICS OF THE BRITISH ARMY

THE POLITICS OF THE
BRITISH ARMY

HEW STRACHAN

CLARENDON PRESS · OXFORD
1997

Oxford University Press, Great Clarendon Street, Oxford OX2 6DP

Oxford New York

Athens Auckland Bangkok Bogota Bombay Buenos Aires
Calcutta Cape Town Dar es Salaam Delhi Florence Hong Kong
Istanbul Karachi Kuala Lumpur Madras Madrid Melbourne
Mexico City Nairobi Paris Singapore Taipei Tokyo Toronto
and associated companies in
Berlin Ibadan

Oxford is a trade mark of Oxford University Press

Published in the United States
by Oxford University Press Inc., New York

British Library Cataloguing in Publication Data
Data available

Library of Congress Cataloging in Publication Data
Strachan. Hew.
The Politics of the British Army / Hew Strachan.
p.   cm.
Includes bibliographical references.
1. Great Britain—History, Military.   2. Great Britain—Politics
and government—1485–   3. Soldiers—Great Britain—Political activity.
4. Great Britain.   Army History.
DA65.S76   1997   355'.00941—dc21   97–2241
ISBN 0–19–820670–4

1 3 5 7 9 10 8 6 4 2

Typeset by J&L Composition Ltd, Filey, North Yorkshire
Printed in Great Britain
on acid-free paper by
Bookcraft Ltd., Midsomer-Norton
Nr. Bath, Somerset

For my parents
Iris and Michael Strachan

through an appeal to nationalism—something a successful commander was better equipped to achieve than was a civilian politician. Thus Martin Kitchen, in *The Silent Dictatorship: The Politics of the German High Command under Hindenburg and Ludendorff, 1916–1918* (1976), could describe the latter as 'Bonapartists'. Another example of transference was the application of the term 'praetorianism', most notably by Amos Perlmutter in *The Military and Politics in Modern Times: On Professionals, Praetorians, and Revolutionary Soldiers* (1977). The original reference was to the praetorian guard, the personal bodyguard of the Roman emperor. The guard acquired the power to make and break imperial governments; hence the use of the more general and contemporary abstraction to denote an army's ability to intervene in politics.

The discussion in Chapter 1 of this book, in revisiting the literature of the 1960s and 1970s, uses some of these words and the ideas which they contain. In doing so it risks reopening old wounds. It sets the pragmatism of the historian against the conceptualization of the political scientist; it seeks out continuities, but in doing so it is shaped more by the characteristics of the past, than by the need for contemporary relevance.

These debates lay dormant in the 1980s. Armies themselves became politically quiescent. Military coups in Latin America or in newly independent African states were either less frequent or else less indicative of what was supposed to be a general trend. Liberal democracy was resurgent. Indeed, with the end of the Cold War, it seemed triumphant. Armies became politically innocent once more.

It was not just for these reasons that military historians deserted the themes of 'war and society'. Successful in their bid for academic acceptance, they no longer needed to situate military history in other contexts. War was a subject sufficient unto itself. Operational history staged a spectacular come-back. The third volume of the recent and massive collaborative exercise in French military history, *Histoire militaire de la France*, directed by one of the pioneers in the social analysis of past armies, André Corvisier, says virtually nothing about that most vexed of subjects, the political history of the French army in the Third Republic. The army's protest that it was professional and therefore it was apolitical is effectively taken at face value. Dreyfus is barely mentioned.

As this book makes clear in relation to the British army, this is not an approach I find convincing. My argument is organized by themes, and is driven forward by the ideas which grow from themes. It is the result as much of those precious moments of gestation walking the dog in the mornings as it is of hours in libraries and archives. I have encumbered it with footnotes but not with a bibliography: the former can convey my most immediate obligations, but the latter could never do justice to the accumulated reading of years. The book rests largely on published work, and in its treatment of more recent events it relies heavily on the press (especially in Chapter 9). Much of its detail therefore demands further research, and the results of that investigation may well challenge its conclusions. If this volume ends up by doing no more than ask some important questions, it will have done its job.

I fear that soldiers will be affronted by the suggestion that they are 'political'. They should not be: to be 'political' is no more and no less part of the human condition than to be 'social' or even 'moral'. I count many soldiers as friends and more as colleagues. Some of them are identified in these pages, some not. They have all had an input to this book. The response to those of their profession who attended the lectures encouraged me to believe that the interpretation I put on their attitudes was not without resonance or devoid of conviction.

I must thank my father, Michael Strachan, himself a former soldier, albeit not a professional one, for reading the entire typescript with characteristic thoroughness, and so eliminating obscurities and sloppy syntax. The outcome is dedicated to him and to my mother, in an inadequate recognition of their love and support.

By convention many male authors conclude their introductions with thanks to their wives, expressed in terms that can often seem trite to the reader removed from their conjugal bliss. In my case, neither convention nor triteness should be taken as evidence of anything other than the profoundest gratitude and love.

H. S.

Contents

I

Norms and Practices

It is almost tautologous to refer to the politics of an army. Armies are self-evidently political institutions. War and the creation of standing armies to engage in war were among the defining characteristics in the emergence of the modern nation-state. Nothing empowers a polity more succinctly than its armed forces; no act more clearly defines its ultimate interests than a decision to fight.

That such truisms should bear iteration is a reflection of the degree to which we have been brainwashed by Clausewitz, or rather by the glosses of others on Clausewitz. Famously, Clausewitz wrote in Book 1, chapter 1 of *On War*, 'war is not a mere act of policy but a true political instrument, a continuation of political activity by other means'.[1] Less famously, he went on: 'war in general, and the commander in any specific instance, is entitled to require that the trend and designs of policy shall not be inconsistent with these means.' The first quotation reflects Clausewitz the theorist, the second Clausewitz the practitioner. Strategy is at once *both* political *and* military. The political objectives in war have to be moderated by what is militarily feasible: the formulation of a practicable plan must be bounded by military capabilities. Thus the notion that politics is the superior and stronger element in the equation distorts the reality.

We have too often confused the normative, what is ideal, with the historical, what is real. Many criticisms of Clausewitz, most recently those by John Keegan in *A History of Warfare* (London, 1993), make exactly this mistake. But, to be fair to Keegan, Clausewitz himself frequently obscures the distinction. *On War*'s fullest exposition of war's relationship to politics occurs not in Book 1 but in Book 8, chapter 6B. The general thrust is normative. 'Thus policy',

Clausewitz wrote, 'converts the overwhelmingly destructive element of war into a mere instrument.'[2] He goes on: 'we can now see that the assertion that a major military development, or the plan for one, should be a matter for *purely military* opinion is unacceptable and can be damaging.'[3] But then—in a typically Clausewitzian recognition of an inherent tension—practice intrudes: 'Nor indeed is it sensible to summon soldiers, as many governments do when they are planning a war, and ask them for *purely military advice*.' Politicians need to be militarily informed, and soldiers—more controversially—need to be politically integrated.

What follows is Clausewitz's advice as to the best composition for a government. His own eighteenth-century roots make this section somewhat ill-developed. Clausewitz's norms were, after all, not very breathtaking to a generation accustomed to the fusion of military and political direction in one man. Neither Frederick the Great nor Napoleon had had to struggle with the precise balance between political and military in government, as they had embodied the supreme direction of both in their own persons. But, if the statesman and the soldier are not combined in one person, then, Clausewitz advised, 'the only sound expedient is to make the commander-in-chief a member of the cabinet, so that the cabinet can share in the major aspects of his activities'.[4]

These quotations are from Michael Howard's and Peter Paret's 1976 translation of *On War*. This was based on the 1833 German edition, which was less ambiguous than later editions as to the role of the commander-in-chief in cabinet deliberations.[5] It is, however, a translation which in its pursuit of clarity has arguably emphasized the normative to the detriment of the historical. It was, after all, published at the height of the Cold War, when nuclear weapons gave Clausewitz's abstract notions of 'absolute war' an immediacy that could be no more than glimmered before 1945, and when the ruthless subordination of the military instrument to political direction had become the only way to rationalize the threat of war in international relations.

For almost 100 years up until 1976 English-speaking readers relied on J. J. Graham's translation of *On War*. Unlike the Howard and Paret edition, this version embodied the revisions of later German texts. It declares that the commander-in-chief should be a member of the cabinet so 'that he may take part in its councils and decisions on important occasions'.[6] Thus the purpose of the arrangement is transmuted. Rather than enable the cabinet to advise the commander-in-chief on military matters, the proposal allows the commander-in-chief to take part in all cabinet decisions, military or not. In Graham's version the norm—that strategy should be subject to political direction—remains clear, and was accepted as such by Clausewitz's English-language readers. But the single specific practical proposal opened the door to a very different approach—to the soldier shaping policy in areas other than those that lie within his professional compass. Norms and practices were potentially at odds. Herein was at least one justification for the accusation of ambiguity so often levelled at Clausewitz, principally by his anglophone audience.

A telling illustration of the effect of Graham's translation is its application in the work of Colonel G. F. R. Henderson, possibly the best known and best regarded of pre-1914 British military critics. Peter Paret has argued that, of all British writers, Henderson entered most fully into Clausewitz's thought.[7] Although Henderson's direct acknowledgements to Clausewitz were comparatively rare, his dicta were permeated with Clausewitzian influences. War, he declared in the supplement to the *Encyclopaedia Britannica* in 1902, 'is a political act, initiated and controlled by the Government'. But in asserting norms Henderson ducked their integration with realities. As a historian as well as a theorist, Henderson knew that the practice was more complicated than the simple repetition of Clausewitzian formulae suggested. 'It is hardly necessary to observe', he wrote further on in the same article, 'that no civilian minister, however deeply he might have studied the art of war, could be expected to solve for himself the strategic problems which come before him.' He quoted Thomas Arnold to the effect that 'There must be a point up to which an unprofessional judgement on a professional subject may not only be competent, but of high authority, although beyond that point it cannot venture without presumption and folly. The distinction seems to lie originally in the difference between the power of doing a thing and that of perceiving whether it is well done or not.'[8] The consequence of this tension between ideals and practice was a contradiction in Henderson's writings that he never seriously attempted to resolve.

Henderson's masterwork was *Stonewall Jackson and the American Civil War*. First published in two volumes in 1898, it was reprinted many times thereafter, not least for the edification of students at the Staff College where Henderson himself was professor of military art and history between 1892 and 1899. The normative and Clausewitzian Henderson is in evidence. Writing of the failure of the Union commander, George McClellan, to act in the winter of 1861–2, Henderson observed that 'a general who obstinately refuses to place himself in accord with the political situation forfeits the confidence of his employers and the cordial support of the Administration'.[9] But when McClellan did advance into Virginia he failed. Henderson then put much of the blame not on McClellan but on Lincoln. The president was not commended for ensuring the political direction of strategy but castigated for interfering in the conduct of war. Henderson went further. He generalized in terms that were at odds with his own normative prescriptions. He observed that the Confederates flourished while Lee exercised supreme command but languished when it was resumed by their president, Jefferson Davis. Conversely the success of the Union was dated by Henderson to March 1864, when Grant was appointed commander-in-chief and Lincoln abdicated his military functions in Grant's favour.[10] Thus the whole balance of the war was made to swing on the virtues of independent military command: unfettered control brought success, political subordination failure.

Much of Henderson's confusion was attributable to the limited definition of strategy to which he and his contemporaries were committed. Strategy for Henderson was a matter of 'combinations made out of the enemy's sight', and was contrasted with tactics, 'that is, . . . manœuvres executed in the enemy's presence'. Strategy, Henderson insisted, 'is an art in itself, an art which none may master by the light of nature, but to which, if he is to attain success, a man must serve a long apprenticeship'. He ridiculed the notion of Lincoln 'poring night after night, when his capital was asleep, over the pages of Jomini and Clausewitz'.[11]

The thrust of Henderson's study of Stonewall Jackson lay with this narrow interpretation of strategy, with what today we would call operations. He gave only the briefest attention as to how to reconcile its dictates with those of the imperatives of political control. In a footnote he cited not Clausewitz but another veteran of the Napoleonic wars, Archduke Charles of Austria. Charles's solution did not necessarily ensure the supremacy of political direction. He acknowledged that 'the strategical design depends, as a rule, upon the decisions of cabinets', and he therefore recommended that either 'the leading statesmen should have correct views of the science of war, or should make up for their ignorance by giving their entire confidence to the man to whom the supreme command of the army is entrusted'.[12]

Henderson's book was about the United States, but his precepts were aimed at Britain. America's failure to entrust strategy to the soldiers was blamed on the faults of the British system to which America was heir. Britain had wasted resources in the Napoleonic wars, salvaged success at Waterloo, but then neglected the implications of that war and committed an under-prepared army to a campaign for which it was not designed in the Crimea in 1854.[13] In a note to the chapter in which this interpretation of British military history is to be found—a note significantly entitled 'the evils of civilian control'—Henderson concluded:

Yet, in face of the fact that only when the commanders have been given a free hand, as was Marlborough in the Low Countries, or Wellington in the Peninsula, has the English army been thoroughly efficient, the opinion is not uncommon in England that members of Parliament and journalists are far more capable of organizing an army than even the most experienced soldier.[14]

When Henderson wrote these words he was a serving officer. The immediate implications of what he had to say in this capacity will be explored more fully in Chapter 4. But Henderson was also a British military historian. His attitude to politicians is as characteristic of the latter as of the former. Typically, both groups portray politics as a dirty, murky business; civilians may direct the affairs of the army, but they do so uncomprehendingly; they can too easily make matters of life and death the puppets of factional and party interest; parsimony tends to prevail over prudent investment. These themes, no more than sketched in by Henderson, were developed and then repeated by the British army's most prolific and most dedicated historian, Sir John Fortescue.

The first volume of Fortescue's *History of the British Army* was published in 1899. Twelve more volumes followed, and a refrain running through all of them is the author's persistent criticism of parliament. For Fortescue the House of Commons had not protected the constitution and the political liberties that that enshrined, but had constantly threatened them. It had endangered the nation through the pursuit of false economy clothed in its opposition to standing armies. In 1914 Fortescue published his Lees Knowles lectures on military history, delivered at the invitation of his old college, Trinity. Speaking of eighteenth-century House of Commons debates, he said: 'It has been my hard fate to wade through a prodigious number of speeches upon this subject and I have been absolutely nauseated by their hollowness and cant.' And, later in the same lecture, when adumbrating on the small size of the British army, he stated: 'Do not think that I am "talking politics". I am only stating plain facts. I cannot discuss, nor even propound, the questions which these facts suggest; but I cannot avoid the assertion of the facts themselves, for they are essential to our understanding of our subject—they are indeed the pith of British military history.'[15]

Fortescue's disclaimer is typical. Of course he was 'talking politics'. In the same year in which his lectures were first published, elements in the army declared their hand over Irish home rule. They refused to take part in the coercion of Ulster should the government order it. Some of the views expressed then confirm the same confused belief in the separateness of army and politics as that which assailed Fortescue. The general officer commanding at Aldershot, Sir Douglas Haig, wrote to Major Philip Howell of the 4th Hussars, on 27 March, that both his divisional and brigade commanders and those to whom he had spoken at the War Office 'All agreed *the Army must have nothing to do with politics*' (Haig's emphasis).[16] On the same day Major A. C. Grant-Duff addressed the officers of 1st Battalion, the Black Watch, also at Aldershot: 'The greatest misfortune which can befall an Army is to be involved in politics. Like the Crown, to whom our allegiance is due, we have nothing to do with politics. Through no fault of its own, so far as we can judge, this disaster has befallen us and the army has been dragged, at any rate for the moment, into the mire of party politics.'[17]

These were of course expressions of the norm. The move to outright condemnation, and thus apparently to a further distancing from political involvement, was a small step. Another Highland officer, Major A. W. F. Baird of the 1st Battalion, Gordon Highlanders, wrote to the Labour party leader, Ramsay Macdonald, on 25 April 1914:

I for one, & hundreds of others of *all* ranks in the service, detest *you* & *all* other politicians of all classes and denominations equally.

In our opinion you are all tarred with the same brush. Your personal interests and your party interests, in so far as the latter coincide with the former (& your £400 a year makes the coincidence wonderfully accurate no doubt!) are all that any of you care for. The King, Empire, & the Flag which are everything to us are little or nothing to any of you.[18]

The man at the heart of the Curragh incident, the general officer commanding in Ireland, Sir Arthur Paget, put Baird's view even more succinctly. He was alleged to have said to his officers: 'Don't you think, officers, that I take orders from these swines of politicians. No, I only take orders from the Sovereign.'[19]

Paget—like Haig, Grant-Duff, and Baird—was of course making comments that were themselves deeply political. If it is not too contradictory to say so, claims to be apolitical from an institution whose function is supremely political are themselves political statements.

Of course, this has not prevented British soldiers stressing, with increasing rather than diminishing frequency, that their army is apolitical. For reasons that are associated with the memory of the New Model Army and Cromwell's major-generals, and are embodied in the legacy of the 1688 revolution and the Bill of Rights, parliamentary sovereignty over the army is assumed to be the norm. This is the conventional wisdom of recent historians. Michael Howard, introducing the volume of essays on *Soldiers and Governments* which he edited in 1957, wrote of the British army in 1914, 'For over two centuries the supremacy of the civil power had been unquestioned. Neither the Army nor the Navy had a political past.'[20] Correlli Barnett, in *Britain and her Army 1509–1970*, described Britain's army as 'militarily powerful, but politically insignificant'.[21] Field Marshal Lord Carver—who as a former chief of the defence staff could certainly lay claim to greater insight on these matters than most—said virtually nothing about the politics of the army or the political status of the army in *The Seven Ages of the British Army*, published in 1984. His book was, essentially, a politically neutral statement. And in the most recent one-volume history, *The Oxford Illustrated History of the British Army*, published in 1994, its editors, David Chandler and Ian Beckett, wrote in their introduction, 'compared with other armies, the British army has been largely apolitical'.[22]

To be fair to Chandler and Beckett, they go on to acknowledge that 'the army has played a political role', and they highlight themes that I shall treat in this book. But they do not then ask themselves whether playing 'a political role' may not ultimately undermine the army's apolitical stance. The thrust of this book is the belief that it does.

Vital to the interpretation which sees the British army as apolitical is the comparative context which Chandler and Beckett cite. To be reductionist, since the restoration of the monarchy in 1660 the British army has not staged a military coup. Many other armies, particularly in third world countries, have. In 1958 about 60 per cent of the world's states were governed directly or indirectly by the military.[23]

One of the earliest and still most influential texts on military intervention in politics is S. E. Finer's *The Man on Horseback*, first published in 1962. Finer described four ascending levels of intervention—influence, blackmail, displacement, and supplantation. A coup falls in the last category and it is not appropriate to Britain. But the army has been implicated in at least one case of displacement, the fall of the last Liberal government in May 1915. Moreover, the emphasis on high drama should not lead us to neglect the lower rungs of Finer's ladder. Armies can express their views through normal constitutional channels, through lobbying and through the press; armies can intimidate the civil authorities through threats of resignation or non-cooperation. The British army has used the first throughout its history; it has not infrequently done the second. If we refocus our gaze—if we stop looking for coups that are not there—and instead consider more subliminal activity, the problem becomes that of too much evidence, not too little.

The key point implicit in Finer is that all armies are inherently predisposed to political activity: what curbs their influence is less their own characteristics and more the context in which they are operating. In countries of what he dubbed 'developed political culture', the civilian authorities, by virtue of longevity, popular acceptance, or constitutional legitimacy, do not—except possibly in time of war—provide the opportunities for armies to exploit their potential for exercising power. On the other hand, if civilian political culture is low, armies naturally spill over to fill the vacuum. Thus in Britain, as in other nations with 'developed political cultures', only influence and blackmail normally remain open as channels for military intervention.

The point can be made—even at the expense of begging questions—by considering it in reverse. Britain's army became a role model for many of the nations that gained independence from its empire. In some cases—such as Ghana or Nigeria—independence did not preclude the continued training of these nations' officers at Britain's military academies. Therefore the norm handed down to the armies of the new states was that of non-intervention. But coups have still taken place. In Pakistan the army's involvement in politics has been persistent; in India, its neighbour and the inheritor of a similar military tradition, the army has not intervened. Any general explanation that looks to factors internal to the armies themselves is inherently less satisfactory than an account that embraces a wider political context. What is particularly appealing about this approach is that it enables us to consider the British army's political behaviour on the basis that it is more like other armies than conventional accounts allow.

Received wisdom might—at the risk of caricature—go something like this. The British army is, unusually for armies, apolitical: that is, it does not mount *coups d'état*. The reasons for this begin with the constitutional developments of the seventeenth century—Charles I's wrangles with parliament over financing the military, Cromwell's major-generals, the 'Glorious Revolution', and the 1689 settlement. But external factors also played their part in permitting the gradual and early growth of constitutionalism: Britain's status as an island freed it from the fear of invasion and so reduced its dependence on the army. The passage of time, and Britain's avoidance of sudden and violent change, have caused the nation and the army to internalize the patterns of behaviour to which these circumstances gave birth. What began as the result of the domestic squabbles and the foreign policies of the seventeenth century became ingrained—practices became norms. This interpretation appeals to the army as much as it does to the nation as a whole: Britain's institutions mark it out as having undergone experiences clearly differentiated from those of the Continental powers.

However, this approach presents us with a paradox. The factors that explain the army's political subordination in the seventeenth century are external to the army; the factors that explain the same behaviour in subsequent centuries are not—or not to the same degree. The argument, therefore, is that the political nature of the army has changed. Indeed it would be surprising if it had not. But has it changed in the way in which the conventional wisdom suggests it has? Has it really been the case that the army has been denatured—that it, unlike other armies, has, exceptionally, lost its political attributes? If it has, the explanations for the apolitical behaviour of the British army are intrinsic rather than extrinsic to the army. But such explanations get us into all sorts of bother.

There are, in essence, two interpretative traditions here, each to a large extent exclusive of the other. One, associated with Samuel Huntington, argues that the more professional an army is, the less likely it is to intervene in politics. The second, advanced originally by Morris Janowitz, puts the opposite case: the greater the professionalism of the military, the more obtrusive it will be in civil–military relations.

Both studies, it must be stated at the outset, are concerned primarily with the United States army. The relevance of Britain's experience to the United States might make it seem as reasonable to draw conclusions about the former from the latter as vice versa. Both, after all, were inheritors of the same seventeenth-century tradition, and the United States has, like Britain, enjoyed comparative immunity from invasion while pursuing its development as a liberal democracy. However, Huntington, writing in 1957, and

Janowitz, writing in 1960, were driven less by this long-term perspective and more by the immediate changes wrought in the American armed services through the Cold War. They were concerned
with America's problems in adapting to the maintenance of a large
military establishment in peacetime, and with the attendant difficulties of social and political integration. Their mastery of Britain's
military history was, especially in the case of Janowitz, limited.

Samuel P. Huntington's *The Soldier and the State*, subtitled 'The
Theory and Politics of Civil–Military Relations', was a path-breaking study. He argued that the more professional an army became,
the more politically compliant it would be. An army's very military
qualities could be the agents in separating it from the political life of
the state. This would be particularly true in countries that had
confronted few threats to their security and therefore had ideologies
that were averse to armies and accorded armies little political
power.[24] An army with norms that were apolitical had, effectively,
conformed to one definition of professionalism.

Britain, despite its constitutional relevance to the United States,
presented problems that Huntington did not fully reconcile. He
argued that professionalism in Britain was retarded by two factors.
First, the dual political control established in the 1689 settlement
generated rivalry between crown and parliament that involved the
army in domestic politics.[25] I agree with much of this: indeed the
struggle for control of the army was one of the prime factors in
fashioning the army as a political entity. The crucial question is
whether such competition necessarily retarded the advent of professionalism. Secondly, the aristocracy retained its hold on commissions in the army through the exercise of purchase. Thus both
professional education and the development of a proper career
structure were held back. Huntington therefore confronts the conundrum that for him the British army, historically speaking, is
neither professional nor—despite his first point—overtly interven-
tionist. He squares his circle by attributing its compliance to the
dominance within the army of the aristocracy: 'direct socio-political
controls were preferred to reliance upon a sense of professional
responsibility.'[26]

Morris Janowitz's sociological analysis of the British army in *The
Professional Soldier: A Social and Political Portrait* is comparable to—and

as crude as—that of Huntington. 'Historically, the officer's social prestige was regulated by his family origin and by an ethos which prized heroism and service to the state. What society at large thought of him was of little importance, as long as his immediate circle recognized his calling. This was particularly true of the British officer corps with its aristocratic and landed-gentry background and its respectable middle-class service families.'

But Janowitz then goes on to point to radically different conclusions concerning professionalism from those espoused by Huntington.

As the military profession grows larger and socially more heterogeneous, as it becomes more of a career, does not pressure develop for prestige recognition by the government at large? Every professional soldier, like every businessman or government official, represents his establishment and must work to enhance the prestige of his profession. In turn, a military figure can become a device for enhancing a civilian enterprise. Do not such trends force the military to become more obtrusive and place a strain on traditional patterns of civilian–military relations?[27]

In the field of military sociology, Janowitz—rather than Huntington—has called most of the shots. A broadening in the social composition of the officer corps has been associated with the advent of professionalization, and this professionalism has not necessarily stood in the path of political intervention. However, this creates difficulties in accounting for what Janowitz sees as the continuing low level of political involvement in the British army. One of Janowitz's disciples, Gwyn Harries-Jenkins—on being asked to contribute a chapter on Britain for a book published in 1983 on the political education of soldiers—could do little more than state his perplexity. The rationale for the book was Janowitzian: professional soldiers need 'civic education' to curb their inherent proclivity to political activity. Harries-Jenkins had to confess that 'In the United Kingdom, we are conventionally presented with a military system which seemingly sustains itself on professionalism and tradition rather than on the development of civic education. Yet such a professionalized military, notwithstanding the caveats of the textbook writers, is in effect apolitical and value-neutral.'[28]

One obvious and logical way out of the Janowitzian dilemma is to revert to external explanations. Freedom from invasion, the early development of a police force, colonial campaigning, a fully legitimized civilian authority—all these are adduced to explain the army's low level of political activity. Janowitz himself—albeit showing in the process an extraordinary blindness to the purposes and pattern of imperial conquest—argues that the army was never thrust into the position of assuming 'that its primary task was to force British standards of behaviour on unfriendly states'. Because it 'was charged with the task of limited objectives', its political horizons were also circumscribed.[29]

However, Janowitz will not let go of the argument that social profile is as important as, if not more important than, the army's functions in determining its inclination for politics. In this respect, Huntington's theory presents its author with fewer problems. Huntington reckons that, despite its slow start, the British army began to be professionalized in the course of the late nineteenth century.[30] Thus, 100 years later, the British army is fully professionalized and therefore it is politically subordinate. Because Janowitz accepts the second part of this proposition, his theory leaves him no choice but to reject the first. His social profile of the British army in earlier epochs has much in common with that given by Huntington. But the corollary is, of course, political neutrality (where for Huntington the absence of professionalism had connotations of political activism). 'Birth, family connections and common values', Janowitz wrote in 1968, 'insured that the military embodied the ideology of the dominant groups in society. Political control was civilian control because there was a unity of interests between aristocratic and military groups.' To explain the continuance of this low level of political involvement Janowitz has to propel its assumptions forward. The pace of change in the British officer corps, the absorption of the middle classes, and the establishment of professional standards—already seen as slow by Huntington—were reduced to snail's pace by Janowitz. Even in the 1960s, he insisted, the educational system and the formal requirements for entry continued to favour the sons of upper-middle-class families.[31]

Janowitz's hypothesis therefore rests on the assumption—in contradistinction to Huntington—that the army escaped the professionalizing influences at work in nineteenth-century Britain. This view was tested in a more extensive study of the officers of the Victorian army written by Gwyn Harries-Jenkins and published in 1977. Harries-Jenkins endorsed Janowitz's interpretation. He argued that the officers of the nineteenth-century army were drawn disproportionately from the aristocracy and the gentry. Landed wealth therefore dominated the army just as it dominated parliament, and so the former was identified with the latter by virtue of their common social and economic interests. Officers stood for parliament because of their status as property-owners, not because of their careers as soldiers. In this interpretation the dynamism and energy of the middle classes remained untapped: social background blocked the advent of functionalism.[32]

Harries-Jenkins's argument is vulnerable to at least three lines of attack. One is to do with the nature of the Victorian army itself: even the evidence which he admits in his own support causes his case to creak, and a more rounded view of the Victorian army—one which considers it in the context of its colonial roles rather than in a purely domestic context—produces a very different picture. This point will be developed in later chapters. Secondly, there is the contradiction inherent in the argument that an army which is officered by the politically aware classes is itself likely to be politically neutral. The more obvious assumption might be the exact opposite. And thirdly, there is the presumption that aristocrats and gentry cannot by definition be professionals—a presumption with which somehow the British army, but no other army, is saddled.

Before we go further, it is worth expanding on these last two observations by putting them in a comparative context. Most military historians would regard the German army as a reasonable exemplar of military professionalism—at least judged by performance—between 1871 and 1945. Huntington has real problems with this, because it is not as politically quiescent as his thesis would allow. But the contrast between actuality and norms, between evidence and theory, is far more striking when the German illustration is applied to Janowitz's case. The German army before 1914 worked harder than the British at preserving its aristocratic credentials: indeed this was embodied in the continued domination of Prussia in the military hierarchy. The small size of the post-1919 army enabled the nobility to reassert its hold on the officer corps in the 1920s. This aristocratic element, although proportionately reduced by the expansion of the officer corps in the late 1930s, still remained sufficiently distinct to form the nucleus of the conservative opposition to Hitler and to be the kernel of the July 1944 bomb plot.[33] The point that emerges much more unassailably from Germany than any dispute about the correlation between professionalism and politicization is that aristocratic recruitment to the officer corps has not necessarily been inimical to the growth of professionalism. Indeed, from the Enlightenment onwards, many European nobilities have striven to retain their traditional monopoly in the bearing of arms by embracing the principles of professional military education.[34]

Military sociologists, caught in the grip of Gaetano Mosca's pioneering work on the ruling class,[35] have been extraordinarily reluctant to give up the idea that there is some equation between the social backgrounds of officers and their professionalization. Military historians on the other hand have been less reluctant to judge professionalism in the light of performance. Combat experience, foreign service, and length of career can all be potent factors in inculcating a sense of *esprit de corps*, and with it a sense of professionalism. Officers who have come to identify with their service may well develop corporate loyalties that override any homogeneity between them and the political élite.

The clash between theoretical prescription and empirical evidence is almost audible in one of the most frequently quoted studies of the British army by a Janowitzian. C. B. Otley took a sample of 330 lieutenant-generals, generals, and field marshals between 1870 and 1959. Superficially his conclusions confirmed the generalization that the army élite was linked to the 'old ruling order'. But on closer consideration Otley observed that the really great families of Britain were not providing officers for the senior ranks of the army. 'Instead it is lesser landowners, senior administrators, soldiers and independent professionals—the "supporting class" which has provided the recruits to the army élite.'[36] Otley noted the decline in propertied backgrounds: fully three-quarters of his sample owned no land. By contrast, 39 per cent of the fathers of his senior officers were themselves soldiers. The army was acquiring the features of a hereditary class. Commenting on the existence of these military families, Otley remarked: 'It is hardly surprising that the army officer corps displayed interest-group characteristics which were distinct from if not hostile to their class affiliations and loyalties. Officers developed intense and specifically "military" loyalties which led them to devalue or even to revile the values and practices of civilian society.'[37]

The conclusions to be drawn from Otley's analysis were corroborated by a study of the period 1965 to 1977 undertaken by R. G. L. von Zugbach. He found that between fourteen and seventeen officers below the rank of colonel were titled; if younger sons of peers were included, this figure rose to between thirty-five and forty. Such low figures hardly warranted describing the army, at least collectively, as aristocratic. But, Zugbach's findings suggested that, although the army was not socially exclusive in terms of birth, it was in terms of education. Eighty-four per cent of generals were educated at public schools, and 22.4 per cent were from three schools only, Eton, Harrow, and Winchester. Furthermore, in this respect the trend was deepening, not diminishing: in 1897 only 37 per cent of senior army officers had been educated at public schools.

Once again we are in danger of being lured towards a conclusion that favours class over function, sociology over practicality. Von Zugbach, like Otley, found evidence that the army was behaving as a self-contained profession. The guide to selectors for the Regular Commissions Board said that candidates should have a genuine interest in, and good knowledge of, military life. Both requirements, though perfectly reasonable, were ones which the sons of servicemen were more likely to meet. Between March 1973 and March 1976 31 per cent of those successful at the Regular Commissions Board were the sons of officers or serving other ranks. Candidates who had been to public schools also tended to do well: the qualities of individual initiative allied to social adaptability sought by the Board were exactly those most likely to be imparted by boarding education. These two trends were not independent but different sides of the same coin. Until the late 1960s prolonged periods of imperial service overseas meant that boarding-schools were the logical choice for officers in the education of their children. Thus the simultaneous emergence of the public schools at both ends of the professional career ladder can be seen as one manifestation of Otley's hereditary class, a symbiosis of function and professionalism rather than a consequence of civilian social values.[38]

Sociologists have remained reluctant to embrace the implications of their own findings. The evidence which Otley marshalled brought him close to shattering the whole theoretical structure on which Janowitz's arguments rested. He stated quite explicitly that his study had not sustained the argument that the social origins of officers were a cause of their political subordination. Furthermore, he had suggested quite clearly that British army officers in the first half of the twentieth century belonged to a profession with its own identity and therefore possessed of its own political agenda. But in the last analysis he backed off. His conclusion perversely reverted to an endorsement of Mosca: class loyalties overrode corporate loyalties.

To be fair to Otley and to Janowitz, both of them stressed that what determined the likelihood of military intervention in politics in general terms was the legitimacy of the civilian government. Britain's government in the twentieth century has been as legitimized as any in the world. Therefore, even if the British army was fully professionalized, and even if that professionalism predisposed the army to political activity, its effects would be limited by the context in which it was operating. This is, effectively, the thesis of the most significant recent interpretative survey of the military and politics, that by Amos Perlmutter, published in 1977.

Perlmutter identifies three types of military organization in the modern nation-state. The classical professional soldier prevails in stable political systems; the praetorian soldier thrives on political instability; the revolutionary soldier is linked to a political order that is stable despite its origins in a declining or new political system. What varies, therefore, is the political order, and it is this which shapes the character of the soldier. The soldier himself is in a job which is inherently political. Perlmutter—endorsing a view already embraced in this chapter and reminiscent of Finer, if not of Huntington—explicitly states that 'the military cannot take a neutral political stance'.[39]

Perlmutter's justification for this view clearly links the growth of professionalism to an increasing political awareness:

The propensity of the military professional to intervene in politics and in policy formation is linked to his corporate and bureaucratic roles and orientations. As a corporate body the military organization strives for internal control of its profession and for protection from external political control . . . Striving to maximize autonomy entails, among other things, exercising influence in politics, both through organizational counterpressures and through political institutions and regimes. As a bureaucratic profession, the military is in politics to the degree that it is a key partner of civilian politicians and bureaucrats in the formation and implementation of national security policy.[40]

Britain clearly belongs to Perlmutter's first category—that is, it has a professional army operating in a stable political system. As a professional army it is likely, at the very least, to be drawn into politics in the manner suggested by Perlmutter. It is likely to seek to express its opinions through 'influence' and 'blackmail'—to revert to the vocabulary of Finer. Perlmutter himself does not draw out the significance of his argument for Britain because his characterization of the British army and his views of its professional evolution are as limited as those of his predecessors in the field. He reiterates the stock generalizations that the British army was aristocratic, and that as a consequence it was slow to gain a professional identity. Much of what he says is frankly contradictory. Although the army was officered by nobles, the social status of the officer was apparently minimized. Although the officer corps was socially exclusive, it was the preserve of 'third sons and aristocratic failures'.[41]

It is a fundamental proposition of this book that the social composition of the army is largely irrelevant to its growth as a professional body. The preoccupation with social origins has helped obscure the effects of other professionalizing influences which began to operate from its first establishment as a standing force after 1660. However, it is also a presumption of this book that Janowitz is right on the inclination of a professional army to intervene in politics. In this respect this book embraces the thinking not only of Janowitz but also of Finer and Perlmutter. Thus the British army has been impelled into politics not least by the very fact of its professionalism.

This is not to impute some sort of malevolent intent to the army. The already-quoted utterances of officers at the time of the Curragh incident in 1914 imply that soldiers who dabble in politics are somehow unclean. It is certainly true that some officers have enjoyed politics, and have thrived on intrigue and conspiracy, more than others. But they are not the exceptions which the holier-than-thou pronouncements of their colleagues might suggest. Rather, they are the cutting edge—the most obvious manifestation of the fact that the army must be politicized because (to quote Perlmutter once again) 'it is in the service of the state and the authorities'.[42] At the risk of yet further repetition, what limits the impact of their intervention is not that the army is inherently apolitical—because it is not—but the political culture within which the army is operating.

2

....

PROFESSIONALS AND POLITICIANS

Even the most charitable of the political scientists and sociologists referred to in the last chapter, Samuel Huntington, would not date the emergence of a professional body of officers in the British army any earlier than the late nineteenth century. For him, as for many others, a principal stumbling block to pushing the origins of professionalism back any earlier is the persistence of the purchase of commissions, which was not finally abolished until 1871.

Purchase is presented as more than just the symbol of aristocratic dominance. It is also seen as a guarantee of political neutrality—a view espoused as fervently by its nineteenth-century defenders as by the latter-day military sociologists. Because promotion depended on purchase, officers were not reliant for preferment on political patronage and therefore did not seek it. Furthermore, because their commissions represented a sizeable financial investment, officers had a property interest in maintaining the political status quo. Revolution could wipe out their investment. Purchase was a block to what Perlmutter and others have called praetorianism.

The proposition that the army became more politically active after the abolition of purchase is arguable but chronologically defensible. However, we should be wary of linking cause and effect. Purchase only operated regimentally, in other words for ranks below that of lieutenant-colonel. In the higher ranks of the army purchase played no part; rather, patronage—moderated by seniority—was vital both before 1871 as well as after. Even at the regimental level, it is possible to exaggerate the role of purchase. Vacancies created by death were not filled by purchase, and in wartime in particular purchase played proportionately little part in sustaining upward mobility in the army. Before 1871 many ambitious but impoverished officers found sufficient avenues to advance their careers despite purchase: certainly any suggestion that purchase in itself precluded praetorianism would be an exaggeration of its domination.[1]

There is a second and even more substantial difficulty with attaching so much salience to purchase. Its abolition did not produce a significant shift in the social composition of the officer corps.[2] The ability to purchase did not, after all, depend on birth: it depended on wealth, and therefore it did not distinguish between old money and new. Indeed it could be argued that it was precisely to counter the possibility of new wealth and its incursions that some of the more traditional officer-producing classes (and that does not mean solely the aristocracy) favoured the abolition of purchase. Younger sons of aristocrats, who were themselves landowners caught in the agricultural depression of the 1870s, or the offspring of the professional classes, dependent as clergymen or even army officers on fixed incomes, had good cause to be grateful for the cessation of purchase.[3]

If purchase is a red herring, if its abolition is indicative less of change than of underlying continuities, then one of the principal points to which the advent of professionalism is fixed is removed. It becomes possible to chart professionalization in relation to practice and performance rather than to sociological determinism. This is not to say that the social composition of the officer corps is entirely irrelevant to what follows, but that is because the army's commissions were never the exclusive preserve of the aristocracy. Even in the eighteenth century, the British aristocracy had neither the numbers nor the inclination to exert the sort of stranglehold on commissions established by the nobilities of France or Prussia. Primogeniture restricted the number of peers, and many found sufficient economic opportunity in agriculture, commerce, and even industry to make the exercise of a monopoly on commissions redundant. Of course, there was a strong correlation between officership, landed wealth, and title, particularly in the senior ranks, but the door to new blood was never completely closed. Breeding coexisted with ambition.[4]

John Childs has been able to site the advent of professionalism in the British army almost two centuries before the abolition of purchase. In 1685 James II and VII inherited from his brother Charles II a small but solid core of officers. James respected the army and gave it status: in just over three years the English establishment expanded from 8,865 men to 34,320. Growth facilitated promotion, and those who were rewarded were not courtiers but men with previous military experience, acquired either in the service of foreign powers or in the new colony of Tangier. In the process the army's officers became divorced from the gentry and progressively more dependent on the crown. But this brought a danger. James demanded political and religious conformity as a condition of military advancement, and those unable to oblige were likely to find their commissions forfeit. James purged the Irish army, leaving less than fifty out of 406 professional officers. Nearly all of those who had lost their jobs had purchased their commissions, and yet they were not compensated for the loss of their investment or of their livelihood. Professionals across the water in England, similarly dependent on their swords for their incomes, became alarmed. They could not afford to sacrifice their military careers on the altar of the king's religious convictions. Therefore, in 1688 a small group, conspicuous among them John Churchill, conspired to welcome William of Orange when he advanced on London from Torbay. Their commitment to their profession shaped their political behaviour.[5]

William's accession to the throne might have precipitated another purge of the officer corps. But it did not. About a third of James's officers followed James, and a further third sought retirement. William could not afford to dispense with what remained. War with France was what had predisposed him to establishing a hold on Britain in the first place: he could not risk further damage to an instrument of such potential value in that struggle. Thus the professionalism born under James was consolidated under William. By dint of almost continuous Continental campaigning between 1689 and 1713 the British army created competence through battle, a process validated by the victories of the duke of Marlborough.

Vital to this sense of careerism was the advent of half-pay. Half-pay had been used by previous monarchs as a device for keeping in reserve those officers rendered temporarily jobless by the reductions and disbandments consequent on peace. The treaty of Ryswick in 1697 could have caused the dispersion of the officer corps forged over the previous decade: half-pay ensured that it did not. Parliament accepted this device on the assumption that it was temporary, a way of covering the gap until officers could be paid off in full. Instead the resumption of hostilities in 1702 resulted in the recall of half-pay officers to active service, and thus half-pay was established as normal practice rather than as an expedient. But half-pay did more than help create a long-service officer corps; it also left officers free for other occupations when not on campaign. It allowed soldiers to be simultaneously professionals and politicians.[6]

John Brewer has characterized eighteenth-century England as a 'fiscal military state'.[7] Engaged in almost continuous war against France, and busily carving out an empire in India and the Americas, the British army had no opportunity to rest on its laurels after 1714. Instead the pressure for professionalization persisted.

George I was anxious to possess a disciplined and centralized officer corps. He found himself thwarted in his efforts to abolish purchase, and had to settle for its regulation. But, during both his reign and that of his successor, the principle was established that officers held their commissions by virtue of their education rather than their wealth. These were men to whom pay mattered—many of them sons of professional fathers, and some of the older subalterns ex-rankers. George I recognized this by extending the working of half-pay so that it not only kept officers in reserve but also provided pensions for those disabled in action or anxious to retire from the service. Symptomatic of the army's professional status is the way in which it was used as a vehicle for advancement by the ambitious. Most noticeable were the numbers of Scots, before but particularly after the union of the parliaments in 1707. Between 1714 and 1763 a quarter of army officers were Scots—proportionally more than were English. They used the army as a way to exploit the opportunities opened out to them by the creation of Britain and by its acquisition of empire.[8]

The Georges responded to this pressure from below by rewarding those who distinguished themselves in the service. The principle of regimental proprietorship, vested in the colonel of the regiment, was the main block to monarchical control, forcing the king—as Alan Guy has pointed out—to focus his centralising instincts on the more junior ranks. But as regimental colonelcies fell vacant, the crown used its patronage in ways that recognized martial achievement and therefore did not always work to the advantage of the aristocracy and the gentry. Of 290 colonels appointed between 1714 and 1763, over twenty had served for more than forty-five years, over sixty had served between thirty-five and forty-five years, and ninety had served over fifteen years.[9] The first two Georges therefore promoted the development of long-service careers. In 1740 it took an average of twenty-one years to reach the rank of lieutenant-colonel in the Horse. J. A. Houlding, surveying the eighteenth century as a whole, has observed the growth of military families, of sons following the careers of their fathers. The pattern observed by C. B. Otley for the early twentieth century, and commented on in the previous chapter, thus found its origins well over 100 years before. The names of these army dynasties—Churchill, Lascelles, Howard, Campbell—can carry aristocratic connotations, but often the title was gained through outstanding service in the army. In this context Marlborough is not the exception but the most distinguished case that proves the rule.[10]

The Napoleonic wars stand as the culmination of this eighteenth-century professionalization. The army was more than six times as big in 1814 as it had been in 1789, and it campaigned almost continuously over two decades. But the momentum thus sustained under Wellington could easily have been lost after 1815. Some would say it was. Post-war contraction and a century without a major Continental war could have had stultifying effects. They did not. As in the eighteenth century, so in the nineteenth the half-pay list created a sense of 'once a soldier, always a soldier'. The spate of military memoirs, largely prompted by Britain's first major work of military history, William Napier's *History of the War in the Peninsula*, helped disseminate a military ethos. But most significant of all for the army was the persistence throughout the nineteenth century of colonial campaigning. Britain was never really at peace between 1815 and 1914, and was frequently engaged simultaneously in two or even three theatres of war. It is absurd to exclude the British army from the professionalizing forces at work in the nineteenth century when its cumulative experience of actual operations must have rivalled, or probably exceeded, that of any other power in the world. Furthermore, the fact that most of this service, at least for the infantry of the line, was passed in climates distant from the attractions of home society made the army an inward-looking institution, a self-contained community driven by physical separation into creating its own patterns of behaviour, its own conventions and recreations.[11]

Those who were not committed to the army as a career did not persist in the service. Gwyn Harries-Jenkins's analysis, in arguing for an identification between the army and domestic political élites in Victoria's reign, makes the mistake of identifying the cavalry and the guards—for both of which the frequency of foreign service was less and aristocratic representation greater—with the army as a whole. Even the figures used by Harries-Jenkins (and originally prepared by P. E. Razzell) suggest that the picture was more complex than his focus on the aristocracy allows. In 1875, Razzell and hence Harries-Jenkins argue, 18 per cent of the officers were aristocrats, 32 per cent were landed gentry, and 50 per cent middle class.[12] The largest single group—at a date too soon after the abolition of purchase for this to have had a significant impact on the profile of the officers as a whole—is therefore categorized as middle class. And yet it is this group which Harries-Jenkins's study neglects. Above all, it obscures the phenomenon of the military family, already noted by Houlding for the eighteenth century and remarked on by Otley for the twentieth. In 1899 almost a quarter of colonels and generals, in other words of the more successful and more career-minded officers, came from service backgrounds. Twelve per cent of them came from the peerage and baronetage, but of course some at least of those titles had been gained through military achievements. In 1899, as in 1875, about half the total came from middle-class backgrounds—or, more properly, from the professions, including the clergy and the army itself.[13]

It is therefore a presumption of this book that for much of the army's history many of its officers can be characterized as professionals. If Huntington were right, this contention would be confirmed by the degree to which soldiers have shunned political participation. The norm, that the British army is apolitical, would of course suggest that officers have eschewed parliamentary life, so confirming Huntington's model. Nothing could be further from the truth. Throughout its history the army has been one of the best-represented occupational groups in parliament. Frequently officers have ranked second only to lawyers. Nobody seems to have argued that lawyers have forfeited their status as professionals by dint of their political proclivities: indeed, the attraction of politics for the legal profession is seen as a logical extension of the inherently political nature of the law. The analogy with the army is not exact, but neither is it irrelevant. Political disengagement is not the necessary corollary of professionalism. Indeed, it is a second presumption of this book that the reverse can apply—that professionalism can, as Finer, Janowitz, and Perlmutter all argue, make for political activity. Thus the practice (as opposed to the norm)—that officers have embraced parliamentary life—does not invalidate the thesis that they are professionalized; it may even endorse it.

One of the most obvious objections to the rule of James II was that he brought army officers into parliament. Often sitting as members for garrison towns, and financially exposed if they lost the king's favour, these professionals were royal placemen, expected to support the court. Under Charles II the maximum number of army officers in the Commons at any time was forty-one, a total achieved in 1678. In 1688 alone James directed thirty-eight officers into parliament, and John Childs has reckoned that, if parliament had met in 1689, it would have contained between sixty and seventy officers. James even set about the militarization of local government; in thirty-eight English and Welsh counties eighty-two army officers were appointed deputy-lieutenants. The king was using the army to create a centralized and militarized monarchy comparable with those to be found on the continent of Europe.[14]

Given the rhetoric of the Glorious Revolution of 1689, it would be reasonable to expect that parliament would be purged of this military presence. It was not. There was a slight but only temporary fall in the army's representation. Between 1660 and 1690 10 per cent of members of parliament were army officers; between 1715 and 1722 this figure slipped to 8 per cent. But as the numbers of MPs declined—from 739 in 1715–22 to 671 by 1747–54—the numbers of officers rose, from fifty-eight in the first period to sixty-eight in the second. Thus the ratio of soldiers in the Commons for most of the eighteenth century ran at levels comparable with those under the later Stuarts.

It was not the Glorious Revolution but the French Revolution which changed the level of military representation in parliament. Between 1790 and 1820 almost 20 per cent of MPs could claim service in the regular army. In 1812 165 MPs held commissions at the time of their return to parliament, and over the thirty-year period as a whole 256 MPs were currently serving with the army at the time of their first election.[15]

Sir Lewis Namier, reflecting an older historiography, as well as anticipating the conclusions of the political scientists, attributed this high proportion of soldiers in part to 'the fact that the army was still in the eighteenth century a very amateurish service'.[16] Certainly the domination of the Foot Guards in the 1790–1820 cohort of MPs— 107 out of just over 400[17]—would suggest a close link between wealth, birth, and military-cum-parliamentary activity. The fact that the guards rarely went abroad and were stationed in and around London meant that military and political duties could be easily combined. But there are other indicators, noted not least by Namier himself, which suggest that, even if in some cases this image of aristocratic amateurishness is warranted, it cannot be applied across the board.[18]

Of 208 army officers in Namier's and Brooke's list of members for 1754 to 1790, fifty-six were Scots. The Scots were under-represented in the Foot Guards, but massively over-represented in the army's senior ranks: 69 per cent became colonels. The ambitious, long-service careerist was therefore as evident in parliament as he was in the army as a whole. The disproportionately high number of Scottish soldiers in parliament continued at a similar rate—almost one in four—in the period 1790 to 1820. It was, however, the sustained nature of the army's campaigning that really made these late eighteenth-century military MPs—whatever their social or national backgrounds—truly professional. Almost 100 of them served in the Peninsula, and twenty-five were at Waterloo. The most famous was Arthur Wellesley himself, but their ranks numbered many of his senior subordinates, including Rowland Hill,

Robert Craufurd, William Carr Beresford, Stapleton Cotton, and Edward Paget.[19]

Under George I, officers in parliament—like their predecessors before 1689—were expected to support the king. In 1715 six MPs or former MPs were dismissed or forced to sell their commissions because of their Tory affiliations, and two years later seven more followed the same route. In 1734 the opposition promoted a bill to make officers below the rank of lieutenant-colonel removable only by court martial or by an address to either house of parliament. It failed. Two years later, William Pitt was deprived of his commission for his views on the marriage of the prince of Wales. Pitt's response to this indignity was to make it a condition of his joining the government in 1745 that the 1742 Place Act be extended to exclude all regimental officers (as well as all naval officers under the rank of captain) from sitting in parliament. Although the prince of Wales promised to promote such a bill, nothing transpired. The belief was therefore widespread that loyalty in parliament was a path to military preferment.

However, the link was exaggerated and its significance progressively declined. The fact that changes had been mooted in 1734 and 1745 was itself indicative of a shift in attitudes. Promotion beyond lieutenant-colonel was by seniority, and royal favour at these senior levels could only influence appointments, not actual rank. Furthermore, the notion that jobs in the army constituted a field for political patronage ran counter to the anxiety of the Georges to centralize and professionalize the officer corps.

In 1744 two officers voted against the Hanoverians but were not dismissed, and in 1749 two more initiated an attack on the duke of Cumberland's performance as captain-general and yet evaded punishment. In 1764 Lieutenant-General Henry Seymour Conway, MP for Thetford, was deprived of the colonelcy of the 1st or Royal Dragoons because he voted against the government. The opposition protested, claiming that a military appointment could only be forfeit for a military offence, not for political behaviour. Although George III did not reinstate Conway as colonel of his regiment, he never again breached the principle enunciated by Conway's supporters. Conway himself went on to be commander-in-chief in England.[20]

Thereafter, and particularly by the 1830s and 1840s, the freedom from royal or political patronage meant that some at least of the soldiers sitting in parliament were of radical persuasion. Two in particular stand out.

Thomas Perronet Thompson served in the Peninsula and India, rising to the rank of lieutenant-colonel. Captivated by Napoleon's brilliance in the field, he was a keen student of military affairs, reading the works of Saxe, Bulow, and Jomini. He himself wrote a pamphlet on cavalry outposts. He was thus a professional. But his profession took him to be governor of Sierra Leone in 1808, and a political agent in Persia in 1819. Schemes for the economic and moral development of these areas constituted part of the process by which he shifted from evangelical Toryism to Benthamite radicalism. In the 1820s and in 1830 in particular his liberalism led him to act as an adviser on military matters to the revolutionaries of Spain, Greece, and elsewhere. In 1829 his father's death gave him the wherewithal to purchase the reforming journal the *Westminster Review*, and in 1836 he was returned to parliament, where he spoke against the Corn Laws and the House of Lords, and in favour of free trade, currency reform, and the secret ballot. Already sufficiently far up the promotion ladder to be dependent solely on seniority for further advancement, he became a full general in 1868.[21]

After he entered parliament, Thompson did not serve again. Sir George de Lacy Evans, another Peninsular veteran, did. Although elected as a radical MP, first for Rye in 1830–1 and then for Westminster in 1833, Evans never lost sight of his first calling. Between 1835 and 1837 he commanded the British Auxiliary Legion in the Carlist wars in Spain. He was convinced that his politics made him *persona non grata* with Wellington, a view perhaps justified by the fact that he next found military employment after the latter's death. In 1853 he was given command of a division at the first camp of exercise held in Britain since Waterloo, at Chobham, and in the following year he again commanded a division, this time in the more testing circumstances of the Crimea. Like many radicals of the 1830s, he found that the pace of change in both financial and political reform left his own position looking increasingly moderate. But this observation was less true of his views on army reform. He advocated the abolition of purchase and of corporal punishment, he believed soldiers should enlist for a fixed and limited period, and he favoured the advancement of officer education. In 1861 Evans was promoted full general, and in 1865 he retired from parliament. His career is an excellent example of professionalism allied to politicization, and of the interacting effects of the two.[22]

The close correlation between political life and the army persisted in the years of Liberal dominance. Of the 456 English Liberal MPs returned between 1859 and 1874, 122 had military links.[23] But the nature of the connection changed during the second half of the nineteenth century.

The expansion of the army in the Napoleonic wars, and its constriction after 1815, left a half-pay list so swollen that in 1831 there were still 9,404 officers drawing half-pay as opposed to 6,768 on full pay. By the 1850s and 1860s death had wiped out most of these veterans, while the burden of colonial service kept their successors, now fewer in number, in much fuller employment. The pressures of army life meant that the opportunities for political activity dwindled. There were still seventy-one MPs with military connections in 1853, but only twenty-five were on the active list.[24] In 1868 there remained just thirty-four MPs who had served or were still serving in the regular army.[25]

It was therefore practice that made easier the assertion of norms. After the Crimean war, the notion of dual control became embedded in the Victorians' sense of constitutional propriety. In theory at least, the army was rendered politically neutral by the division in its management: the crown was responsible for its command and parliament for its funding. The ramifications of this thesis will be explored in the next chapter. The significant point here is that it made possible the argument that an officer, as a subordinate of the crown, could not also be a member of parliament, and thus a servant of the people. Legally this was nonsense. Even Charles Clode, who in 1869 wrote a definitive constitutional history of the army, clearly stated that a soldier required no permission from the crown to leave his regiment so as to attend parliament: indeed it was his duty to go. Furthermore, the soldier had as much right as any other citizen to express his political opinions. The problem was one of practicalities, not of law. Expressions of political opinion—

not only in parliament but in the press or in public assemblies—could be prejudicial to the internal order of the army. A standing army without discipline would (and Clode quoted Sir Robert Peel at this point) 'be in truth a curse'.[26]

In December 1880, the secretary of state for war, Hugh Childers, was given notice of a question concerning the position of an officer of the Royal Artillery, Major Nolan, MP for Galway, who although drawing full pay had been appointed a Conservative whip. Herein was an opportunity for a full exploration of the ambiguities highlighted by Clode—of the mixture of practice, constitutional rights, and political norms. Childers himself leaned in favour of a select committee, and so did the clerk to the House of Commons, Sir Thomas Erskine May. But the speaker, the Rt. Hon. Henry Brand, preferred to address the matter as a purely practical issue—the problem of how a man could be in two places at once. Brand's own solution was that officers on full pay who were elected should be seconded from the service. However, he feared that even this would raise an issue of principle. If the crown stopped an officer's pay while he was an MP, could it be construed as penalizing the people's choice? When the question concerning Nolan was duly put in the Commons on 10 January 1881 its focus was entirely untheoretical, concentrating on duties, discipline, and the proper performance of both sets of obligations. Childers replied in similar terms, making only a cursory reference to the possibility of a select committee. He confessed that, 'from a purely military point of view, his [Nolan's] presence in the House is doubtless inconvenient, as the battery which he is commanding is on its way to India. But I express no opinion whether the principle of *cedant arma togae* ought to be applied in such cases.'[27]

Ironically, the most obvious immediate casualty of this exchange was a supporter of Childers on the issue of regimental reorganization, a leading military critic, and a Liberal, Sir Henry Havelock-Allan, VC. Havelock-Allan was the scion of a professional military family, his father, Sir Henry Havelock, having joined the pantheon of Victorian heroes by relieving Lucknow and dying in the process. In April 1881 Havelock-Allan resigned his seat to take command of an infantry brigade. Ostensibly this was clear evidence of the effectiveness of the speaker's direction. But in 1885 Havelock-Allan was back in parliament as the representative for Durham, and he was still an MP when he was killed on the north-west frontier of India in 1897.

Therefore, if these were norms, they operated only because they were given force by, and interlocked with, the realities of late Victorian soldiering. But these pressures, while they may have conspired to keep regulars away from parliament, also fostered the emergence of a substitute military presence. The fact that Britain's regulars tended to be abroad, so leaving the country's home defences exposed, stimulated the growth of the auxiliary forces—the militia, yeomanry, and volunteers.

Between 1790 and 1820 the threat of French invasion meant that nearly 1,000 MPs, or almost half the members returned over the thirty years, served as part-time soldiers.[28] A second wave of volunteering, again in response to the fear of French invasion, swept the country in 1859–60. In 1861 142 MPs had connections with the auxiliary forces as a whole, and by 1869 130 with the volunteers alone.[29] Thus even when colonial service was drawing the British army overseas and away from Westminster, so depriving its officers of the leisure or opportunity for politics, the collective military interest in parliament remained buoyant.

In the twentieth century the effects of the two world wars, and of the persistence of conscription after the second, were to revitalize the link between military service and a parliamentary career. When war broke out in 1914, many MPs with military backgrounds were eager to join up. The separate spheres to which the speaker had alluded in 1880 became even more confused in practice. A field marshal, Lord Kitchener, was appointed secretary of state for war; a former secretary of state for war, J. E. B. Seely, went to command the Canadian cavalry corps. Winston Churchill began the war as first lord of the admiralty and ended it as minister of munitions, but in the interim contrived to command the 6th Battalion of the Royal Scots Fusiliers at the front. Significantly, Seely and Churchill, both veterans of the South African war, had pursued their political careers before 1914 while sustaining their military credentials by serving in the yeomanry. Perhaps most remarkable of all—if rarely remarked upon—was the case of Lieutenant-General Sir Aylmer Hunter-Weston. Most of those MPs who served as soldiers during the war were—before the war broke out—primarily politicians, or had abandoned the army with that end in view. But Hunter-Weston was first elected in October 1916, as Unionist MP for South Ayrshire, while commanding VIII Corps in the battle of the Somme. At the end of the war the adjutant-general, Sir Nevil Macready, gave vent to his exasperation. He argued that in war, as in peace, an officer who was elected an MP should, during the duration of his service in parliament, either be seconded by his unit or placed on half-pay.[30]

The problem with Macready's prescription was that war service proved a potent attraction to constituency selection committees. In 1918 12 per cent of newly elected Conservative MPs had served in the forces,[31] and between 1919 and 1939 regular officers ranked second only to lawyers as the largest occupational group in the Commons.[32] Of course many of these military MPs were young men who, although swept up in the events of 1914–18, possessed no long-term military ambitions. That said, service as infantry subalterns profoundly affected a generation of politicians that included three prime ministers—Clement Attlee, Anthony Eden, and Harold Macmillan. Indeed, by adding Churchill, it is possible to say that for twenty-three years—between 1940 and 1963—combat experience on the western front provided a common denominator in the premiership.

Equally significant, however, was the fact that parliament claimed the services of men who were clearly career soldiers and who had reached the apex of their profession. Two of the key figures in the Curragh incident of 1914, Sir Henry Wilson and Sir Hubert Gough, presented themselves as candidates. Wilson, who ended the war as chief of the imperial general staff, was elected as a Unionist for North Down in 1921. Gough, who had been the ill-fated commander of the 5th Army in March 1918, was persuaded to stand as an Independent Liberal in Chertsey in 1922.[33] Gough did not get in, but his motive for political intervention—his opposition to the vindictive terms of the Versailles treaty, and his anxiety to reintegrate Germany in the community of nations—was one that found widespread support among other generals. Both Sir Ian Hamilton, the commander-in-chief at Gallipoli, and Sir Frederick Maurice, the director of military operations between 1915 and 1918, used their involvement in the British Legion to build bridges to German veterans.

The Second World War sustained these connections between army and parliament. Members, as before, proved anxious to don uniform, which they then wore in the House. And, again, they were prepared to use information gained in the course of their military service to lobby ministers, so bypassing the normal hierarchy of service communication.[34] When the election was called in 1945, candidates appeared on the hustings still attired in their battledress, thus indirectly legitimizing their political positions. Perhaps the most notable was Denis Healey, an erstwhile member of the Communist party and now a major, who was elected as the Labour representative for Pudsey and Otley. But it was in the Conservative party that a 'good war' provided significant fillips to political advancement. Furthermore its influence spanned not just the political generation of Edward Heath and Enoch Powell, but extended—through Lord Carrington and Willie Whitelaw—into Thatcher's Britain.

Thus, although the overall numbers of officers declined in the 1945 parliament, they remained the second largest occupational group. In 1951 there were forty-eight regular officers in parliament. Most of them were majors or colonels who had retired after more than twenty years' service.[35] To them should be added the large number holding temporary or Territorial Army commissions. In 1962 Philip Abrams calculated that 9 per cent of MPs were former regular officers, and that military representation in the House of Commons was nearly 100 times as great as the officers' representation in society as a whole.[36]

The fact that the army did not sustain severe cuts until the end of the 1950s, and that conscription survived until 1961, helped keep the army's representation in parliament at levels on a par with those common under the later Stuarts and the Hanoverians. Furthermore, soldiering is a young man's occupation, politics less so. Thus the after-effects of conscription lingered on the political scene much longer than they did on the military. All the secretaries of state for defence in the Conservative administrations of Margaret Thatcher could lay claim to time in the army. The first incumbent, Francis Pym, served with the 9th Lancers in the Second World War.

His successor in 1981, John Nott, was in the 2nd Gurkha Rifles from 1952 until 1956. Michael Heseltine, minister from 1983 to 1986, had perhaps the least distinguished military career, even if it was in a prestigious regiment, the Welsh Guards. George Younger saw action in Korea with the Argyll and Sutherland Highlanders, and thereafter continued his connection with the regiment through the Territorial Army. In 1989 Younger was followed by Tom King, who did his national service with the Somerset Light Infantry and the King's African Rifles.

The appointment of a lawyer, Malcolm Rifkind, as King's successor in 1992 was indicative of a more general decline. Less than twenty of the 651 MPs returned to parliament in that year had military experience.[37] Furthermore, in the short term the decline in military representation in the Commons looks set to continue. This is not just the product of the lapse of time since conscription or of the dwindling size of the army. It is also the consequence of increasing Conservative unpopularity at the polls. Regular soldiers have found greater favour with Conservative selection committees than with Labour. In 1974 an officer of the Queen's Dragoon Guards, Lieutenant-Colonel Michael Mates, was selected for the safe Conservative seat of Petersfield while still serving with his regiment.

None the less, the point has frequently been made that, even if army officers are conservative (with a small 'c') in their political instincts and in their social behaviour, they are not a Conservative (with a capital 'C') lobby group. Indeed what is striking is that the significant military presence in parliament has not been associated with any particular party since its separation from the need to provide political support for the crown during the course of the eighteenth century. Furthermore the very existence of such a group has been covert rather than overt, because officers have only rarely coalesced to form a cohesive and distinctive parliamentary body.

None the less, the bipartisan nature of the officer, and even of the officer in parliament, should not necessarily be taken as evidence of political neutrality or even-handedness. Rather, it is a reflection on the nature of parliamentary debate on defence matters. Security policy has, more often than not, been a focus for inter-party consensus rather than for deep division. Moreover, since 1945 the idea of parliamentary control of defence has become increasingly a fiction: the key decisions have been made elsewhere, and have normally been so packaged by the time they have reached the Commons that opposition can seem naïve and serious debate can be rendered redundant. The opportunities for military MPs to show their service colours have therefore been comparatively rare. On the other hand, when issues of professional military importance have arisen in parliament, then the existence of a military lobby which has otherwise lain dormant has been revealed. Put another way round, soldiers have behaved like other citizens when the issues that confront them do not have a professional bearing, but have behaved as soldiers when they do.

The principal focus of this military lobby has been the House of Lords rather than the House of Commons. It was already common by the eighteenth century to reward distinguished service—or in other words professional success—with a peerage. Between 1714 and 1763 thirty-nine colonels of regiments sat in the upper house. Many of them held their titles by virtue of inheritance rather than of service, but the latter category—embracing as it did Marlborough, Cadogan, Ligonier, Amherst, and Sackville—was not inconsiderable.[38] The trend became more evident after 1815. Sir Robert Peel created only five peers while he was prime minister between 1841 and 1846 but three of them—Keane, Hardinge and Gough—were soldiers. In 1853 15 per cent of peers had army connections, by 1898 35 per cent, and by 1910 43 per cent.[39] The two world wars ensured that there was no slackening of the pace in the first half of the twentieth century. Of 556 new peers created between 1901 and 1957, fifty-four were from the three armed services, and collectively they formed after the law the largest single occupational group so honoured.[40]

It was therefore the House of Lords that became the focus for the main debates on the army between 1868 and 1911. The significance of these dates is that the first marks the formation of Gladstone's first Liberal government—arguably *the* Liberal reforming ministry of the century—and the second signifies the showdown between Britain's last Liberal government and the as-yet unreformed House of Lords. The intervening years were scattered with episodes where the upper house rejected or obstructed legislation coming up from the lower house. These clashes are normally interpreted by political historians as the consequence of the failure to reform the House of Lords either in 1832 or subsequently, and are taken as a reflection of the fact that the upper house was never sufficiently flushed with new blood to prevent it becoming a bastion of Conservatism. When the Conservatives were in power, the Lords were docile, passing legislation that if generated by a Liberal leader they would have rejected. When the Liberals were in power, the Lords behaved as the focus for the opposition, guided by party feeling and eschewing their tasks of moderation and revision.

In the case of army reform, however, this interpretation will only take us so far. The issues with which the Liberals grappled had professional repercussions. The Lords contained several distinguished soldiers and provided them with a platform from which to express their views. Most notable among them was the duke of Cambridge, commander-in-chief from 1856 to 1895, a man who felt himself disqualified from debate neither by his office nor by his royal blood. By the end of his career the duke was seen as the embodiment of reaction, in the army as much as in the Lords. Therefore, it would be easy to use him as a peg on which to hang both Conservatives and soldiers, and to suggest some measure of identification between the two. It would also be wrong. Such a device would obscure the differences of opinion on professional issues between the duke of Cambridge and other ennobled officers, and it would also gloss over the fact that both he and they were driven by professional rather than by political motivations. Ironically enough—despite all that Janowitz, Harries-Jenkins, and others have written—the most obvious manifestations of professionalism in a political sense came from the seat of the aristocracy. The army's association with the Victorian political élite bears an interpretation that is precisely the opposite of that which it has had to bear hitherto: rather than silence the military it gave it a mouthpiece—and, in some hands, a megaphone.

The most powerful illustration of these points is the abolition of purchase itself. In 1870 Gladstone's secretary of state for war, Edward Cardwell, concluded that he should end the purchase of commissions. The bill was mauled in the Commons, not least by the lobbying of retired army officers, but passed on its third reading by fifty-eight votes. Its passage in the Lords seemed to depend on the attitude of the duke of Cambridge: the duke did not like the bill but was persuaded not to say so. However, he did not really commend it either. It was defeated on its second reading. The government had to proceed with abolition by royal warrant.

Ostensibly this story confirms the conventional picture: officers opposed reform, the House of Lords collectively did the same, and in conjunction they blocked a measure designed to advance the professionalization of the army. Undoubtedly, there were those who subscribed to one or more of these positions. But they gained weight because they found themselves supported by more radical thinkers who saw Cardwell's measure as inadequate.

Cardwell had not entered office as an abolitionist. His aim had been the control of over-regulation payments for commissions, and he had only gradually come to the conclusion that the best way to achieve this was by ending purchase entirely and by buying out the existing interests of officers. The problem with this approach was that it did not tackle the other key issues which were also regulated by purchase—the current of promotion and the provision for retirement. Before the Liberals had come to power, the Conservatives had already committed themselves to the reform of the purchase system. Their opposition to Cardwell's measure was determined, therefore, not by some blind enthusiasm for an outdated relic or by simple party pressure, but by the sense that what was proposed was inadequate. Disraeli wanted more reform, not less. Progressive army officers in both houses were therefore caught in a double-bind. Should they support the government on the good professional ground that here at least was a beginning to army reform? Or should they oppose it—on the equally good professional ground that what was proposed was inadequate since it did not constitute a full scheme of appointment, promotion, and retirement? Either way, the important division was not that between the advocates of reform and its opponents, but between the advocates of partial reform and total reform.[41]

Even the dilemmas of the duke of Cambridge illustrate these points. He claimed that in theory he favoured abolition, but that he feared that, if purchase went, seniority would determine the current of promotion and so would block the advancement of youth. Whether motivated by this quite proper concern or—as his critics would have it—by mere prejudice, his arguments were consonant with the reservations of those who wanted more radical change. Moreover, his objections showed the contradictory requirements of professionalism and of political neutrality. He was unhappy with Cardwell's bill on professional grounds, and yet both Cardwell and, in particular, Gladstone told him that it was his professional duty as commander-in-chief to support the military policy of the government of the day. If he stuck to his own sense of what his profession demanded of him—as those other soldiers who opposed Cardwell's bill in parliament did—then his defiance would be interpreted not in terms of the army's needs but in terms of its political impact, of Conservative opposition to a Liberal government. The fact that the duke pursued a compromise course—letting the government know the grounds for his doubts but giving it his backing (however lukewarm) in the Lords—suggests that he actually worked quite hard to reconcile the irreconcilable and so honour his own sense of professionalism.[42]

By the time the military lobby in the House of Lords inflicted its next major defeat on the Liberals, in 1909, the distrust between the profession and the party had deepened.

R. B. Haldane is frequently ranked with Cardwell as a Liberal secretary of state for war who was also a great reformer. When he took over the War Office in 1906, he inherited from his Conservative predecessors a succession of failed schemes for the revitalization of the auxiliary forces. Although the volunteer MPs in the Commons had played their part in the unravelling of these proposals, much more powerful were the colonels of the militia. With a longer history and county associations, the militia was rooted in the Lords rather than the Commons, and could include among its senior officers Conservative cabinet ministers such as Selborne and Salisbury.

Haldane wanted to use the militia as a draft-finding force for regular troops serving overseas. Such a scheme commended itself to the Conservative front bench, which was particularly worried by the Russian threat to India and the need to reinforce the garrison on the north-west frontier. But it could not command the support of the militia colonels, who saw the integrity of their units, and thus their role in cementing rural paternalism, as forfeit to the strategic needs of the empire. The opposition of the militia colonels promised to prevent Haldane's scheme from getting through the House of Lords. The upshot was a compromise brokered in large part by the Conservative leader, A. J. Balfour. The militia was given a dual role. Most of its units were to find drafts in wartime, but they would also be eligible for service intact. The first task satisfied the leaders of the opposition, as well as professional opinion; the second quietened the militia colonels in the Lords. The consequence was that the militia was separate from, not integral to, Haldane's new structure for the volunteer forces, the Territorial Army.[43]

Haldane's success was dependent on his adoption of policies for the army that could commend themselves both to the Conservative party and to the military profession. Many generals, and particularly the younger generation, held Haldane in high regard. But this was not an enthusiasm that extended to the Liberal party as a whole. The army's identification with empire caused it to march to a different tune from that played by Gladstonian 'little Englanders'. The belief that the Liberals were soft on imperial defence and were prejudicing the safety of the empire, evident in clashes from the 1880s onwards, will be discussed in Chapter 5. In a domestic context, it found its principal embodiment in opposition to the Liberals' commitment to Irish home rule. The futures of both Ireland and empire provided the undertow in the final confrontation between the Liberal government and the military lobby in the House of Lords.

In 1909 Lloyd George, the chancellor of the exchequer, introduced his 'people's budget'. In it he increased death duties, introduced surtax, and imposed a levy on land values. His rhetoric emphasized the idea that a bunch of idle, greedy, and parasitical landowners were to be milked to pay for the introduction of national insurance and social welfare. However, the real pressure on the budget was the naval estimates. Lloyd George could therefore have presented himself as the saviour of national defence, and so perhaps obviated any clash with the House of Lords. He opted not to do so. In November 1909, the Lords rejected his budget by 350 votes to 75.

In January 1910 the country went to the polls, effectively called upon to take part in a referendum on the budget and on the future of the upper house. The Liberals lost 100 seats, but remained in power. They reintroduced their budget and it passed. They then sponsored a parliament bill designed to prevent the House of Lords from amending financial legislation and to restrict its right to reject other legislation to two successive sessions. The implication of the bill was that a Liberal majority in the Commons would be able to carry home rule for Ireland despite the opposition of the upper house. The Lords threw out this bill in April. A further election in December resulted in another Liberal victory. In May 1911 the Commons passed the parliament bill but the Lords savaged it in committee. Asquith now revealed that the king had undertaken to create sufficient new peers to ensure that the Lords pass the bill. The effect was to divide the Conservatives, some persuaded by their leaders to pass the bill, some opting to abstain, and some persisting in their opposition. The bill passed, but it is this latter group—those consistent in their hostility to the parliament bill, the so-called 'Diehards'—who command our attention.

Some of the 'Diehards' deserve the backwoods connotations of the title. But many do not. After all, if they had got their way, the consequence would have been a mass creation of peers which would have changed the social character of the House of Lords once and for all. The 'Diehards' were comparatively youthful: 47.4 per cent of them were born after 1860 as opposed to 32.2 per cent of the other peers. They had therefore matured as Britain's sense of empire began to reflect an awareness of its underlying fragility. The travails of the army in the South African war suggested to many that the nation was decadent, physically unfit, and poorly educated. The movements which sprang from the defeats of 1899—national efficiency and tariff reform—were integral to imperial defence and to a conviction that the empire must be reinvigorated or it must go down. The 'Diehards' therefore represented less a declining order built on land, old money, and inherited privilege, and more an embryonic radical right, dissatisfied with the Unionists as much as with the Liberals, and motivated by a sense of nationalism that was almost demagogic.

In these circumstances it is perhaps not surprising that 72.3 per cent of the 'Diehards' could claim military or naval service as opposed to 43.4 per cent of the rest of the upper house. Part of their worry was indeed the navy, and Lloyd George was therefore wise to eschew any reference to the Liberals' efforts on behalf of that service when he was seeking the initial confrontation. But only four 'Diehards' were sailors. Seventy-seven, on the other hand, had served in the army. Two professional issues preoccupied them. One of these was regressive: the old county order had been affronted by Haldane's reform of the Territorial Army. Numerically, however, the auxiliary forces—although proportionately more strongly represented among the 'Diehards' than amongst the other peers (50 per cent to 45.9 per cent)—were less important than the regulars. Here the issue was much more radical, that of conscription. In 1902 a National Service League was formed, with cross-party involvement, to call for compulsory military training. Its early existence was unspectacular, but in 1905 the doyen of imperial soldiers, Lord Roberts, became its president. Roberts had only just stepped down as the commander-in-chief of the British army. He was no figurehead. He campaigned tirelessly. He objected to the reform of the Territorial Army not because of its implications for county society but because it undermined the case for conscription. In 1909 Roberts promoted a national service bill; in 1911 Roberts was a pivotal figure among the 'Diehards'.[44]

The Lords had therefore arrogated to itself a sense of responsibility for national defence. Even today this is where the really authoritative voice of service wisdom can be heard, expressed in a political environment. Chiefs of defence staff are nearly always rewarded with peerages. Frequently the opinions which they then voice are at odds with those of the government of the day, and equally frequently they may disagree with each other on professional grounds. But when Lords Hill-Norton, Carver, Lewin, Bramall, or Craig have elected to use their membership of the upper house to give not only their peers but also the public the benefit of their professional wisdom, they have not been called to order on the grounds that they hold the most senior ranks in their respective services and that their professionalism demands political neutrality.

At this level, therefore, it is recognized that professionalism can require a political response, and that the man who gives it does not in the process impugn his professional integrity. The interpenetration of the services and of parliament which these retired professional heads represent is a continuity in the history of the army. It gives the lie to any neat demarcation between the two spheres. Fusion between soldier and politician has been more common than separation. Norms should not be allowed to obscure realities. Professionalism has been a powerful precipitant of politicization.

3

CROWN AND PARLIAMENT: THE POLITICS OF DUAL CONTROL

In 1869 a War Office official, Charles M. Clode, published a weighty, two-volume work, *The Military Forces of the Crown: Their Administration and Government*. In his introduction, Clode explained that the constitutional history of the British army (and that was what he was writing) began in 1688. By then England had experienced 'the evil of two systems,—of an Army of Plebeians exclusively under the Parliament [in other words the New Model Army], and of an Army of Cavaliers exclusively under the Crown [the army of James II]'. In 1688 James was expelled and William III installed in the so-called Glorious Revolution. The subsequent settlement provided, Clode went on, the constitutional safeguards 'for the freedom of the people against the possible adverse action in time of peace of a standing army'. It did this by vesting the command of the army in military officers responsible to the crown, and by vesting the administration of the army in civil ministers responsible to parliament. Thus the army was subordinated to a form of dual control, where one authority could act as a check on the other.[1]

Herein is the orthodoxy. Clode cited with approval the speech of James Mackintosh on the occasion of the 1816 debate on the army estimates. 'The revolution', Mackintosh declared,

found standing armies kept up without the sanction of parliament, and in defiance of law;—the revolution found the king claiming and exercising the power of keeping up as large an army as he could find means (abroad or at home) to pay;—the revolution branded the usurpation, and expelled the usurping king;—the revolution bent the neck of the military power under the yoke of law, and rendered armies the creatures of parliament, made and destroyed by its breath;—it permitted, indeed, the annual vote of an army with an annual grant of money for its support; and an annual mutiny bill for its government rendering that army every year a new establishment to be proposed on particular grounds, and adopted only on the same principle, as if each year were the first of its proposition.[2]

Mackintosh was a lawyer and a Whig. He was also an aspirant historian. He planned a history of England that would begin in 1688, and so would exalt the benefits that the Glorious Revolution had conferred on the country's constitution. In the event lethargy intervened and his history got no further than the revolution itself. The story of the settlement was left to another and greater Whig historian, Lord Macaulay.

Clode referred to Macaulay too. When dealing with the army, Mackintosh had the benefit of inheritance: his father had been a Scots officer. Macaulay's advantage was the wisdom of experience: he was secretary at war between 1839 and 1841. Macaulay was quite clear as to the achievements of the 1689 settlement. When William ascended the throne, the Whigs were averse to standing armies. They saw them as the symbol of monarchical power, and preferred to put their faith in the militia, which was rooted in the county gentry. But the militia carried a danger even greater than domestic despotism: that of defeat on the battlefield. It was militarily inferior to the standing armies of the Continent. After 1689 the Whigs learnt the error of their ways: 'it was proved by experience that, in a well constituted society, professional soldiers may be terrible to a foreign enemy, and yet submissive to the civil power.'[3]

Historiographically, Macaulay was on thin ice—the 1688 rebellion, whose results he so extolled, was made possible by the disobedience of professional soldiers, which he so abhorred. He fudged the issue by seeing Churchill and his fellow conspirators not as frustrated careerists but as representatives of the nation. 'The Revolution, by altering the relative position of the Sovereign and the Parliament, had altered also the relative position of the army and the nation.'[4] What followed was the acceptance of the standing army within a Whig settlement:

The old national antipathy to permanent military establishments, an antipathy which was once reasonable and salutary, but which lasted some time after it had become unreasonable and noxious, has gradually yielded to the irresistible force of circumstances. We have made the discovery, that an army may be constituted as to be in the highest degree efficient against an enemy, and yet obsequious to the civil magistrate. We have long ceased to apprehend danger to law and to freedom from the license of troops, and from the ambition of victorious generals. An alarmist who should now talk such language as was common five generations ago, who should call for the entire disbanding of the land force of the realm, and who should gravely predict that the warriors of Inkerman and Delhi would depose the Queen, dissolve the Parliament, and plunder the Bank, would be regarded as fit only for a cell in Saint Luke's.[5]

Here was the Whig view of history. Up to a point, it was even correct.

The 1689 settlement was indeed built upon the arguments developed in opposition to standing armies in the big debates of the seventeenth century. The Petition of Right in 1628 had—in condemning the levying of taxes, the billeting of soldiers, and the use of martial law—challenged the king's prerogative to raise troops without parliamentary consent. In 1641–2 parliament had tried to wrest the command of the militia from the king. In a very real sense, therefore, the issues concerning the raising and management of troops had been a precipitant of the Civil War. That war saw the emergence of Britain's first drilled, disciplined, and regimented army, a symbol of the arrival of the 'military revolution' in England. But the New Model Army was not only professional, it was also deeply politicized: its radicalism meant that it 'became an instrument to secure a revolutionary government whose base of popular support grew increasingly narrow'.[6] In 1655 Cromwell tried to impose a measure of central control, as well as of good order, through the division of England into eleven districts, each administered by a major-general. In creating a militia which was subordinate to the major-generals rather than to the county gentry, Cromwell directed a further blow at local management. However, the new militia had at least one merit in the eyes of Cromwell's critics—it was a counterweight to the New Model Army. One of the attractions in the restoration of the monarchy in 1660 was that it involved the disbandment of the New Model Army. This is not to say that Charles II felt that he could do without troops, but he used other methods to raise them. First he established a force of personal guards which by previous standards could be deemed larger than was strictly necessary. And, secondly, he hoped to emulate Cromwell in incorporating the militia under his control rather than that of the county gentry.

Lois G. Schwoerer, the historian of these debates, and of the ideology which they generated, concludes that by 1660 four main themes had become evident. First, parliament and not the executive should command any military force, whether it was a standing army or a militia. Secondly, the military authority of the state over the individual should be limited. Thirdly, army officers should be barred from the legislature because of the conflict of interest inherent in their position. And, fourthly, the militia should serve as the counterpoise to the army, representing the interests of the local and parliamentary gentry.[7]

These arguments, although developed in the context of the specific crises of the seventeenth century, continued to be deployed in eighteenth-century discussions. They moved from the particular to the universal in their application. Although ostensibly they focused on the power of the army, the fundamental agenda was the power of the king. Their thrust was to forestall the creation of an absolutist militarized monarchy of the sort developing in France or Prussia, and prefigured in England under Cromwell and James II.

In this general objective—that is, in curbing the executive powers of the monarchy—the anti-army rhetoric of the seventeenth and eighteenth centuries was, broadly speaking, successful. Britain was both militant and militarized, but its constitution developed in directions that were largely liberal and libertarian. However, in its particulars the ideology was almost a complete failure. None of the four major themes identified by Schwoerer was given practical effect.

First, the command of the army was vested not in parliament but in the executive. Secondly, it was true that the military authority of the state over the individual was limited in the sense that conscription was not—at least formally—recognized. However, once in the army, the soldier discovered that he did not have the same civic rights as others, that certain freedoms were forfeit to military law and army discipline. Thirdly, army officers—as the previous chapter has revealed at length—were not debarred from the legislature. Fourthly, and most surprisingly, the militia—the so-called 'constitutional' force—actually went into decline after 1689, and increasingly became a supplement to the regular army, not a counterweight to it.[8]

Thus the 1689 settlement delivered none of the things, at least in regards to the army, that its ideologues demanded of it. So, what did it do with respect to England's military arrangements? How could it justify Mackintosh's and Macaulay's self-satisfaction? Its provisions rested on two elements, the Bill of Rights and the Mutiny Act.

Outwardly the Bill of Rights was a clear reflection of the fears of those who had lived through the New Model Army's bossing of parliament and who had witnessed James II's efforts to render his army an instrument of absolutism. It declared that 'the raising or keeping of a standing army within the kingdom in time of peace, unless it be with the consent of Parliament, is against the Law'. This led the great jurist Sir William Blackstone to declare in his *Commentaries on the Laws of England* (1765–9) that the laws of England knew no such state as a perpetual standing army.[9] Many, including Mackintosh, have followed in these distinguished footsteps. However, Blackstone was wrong—arguably on the legal point and certainly on the practice. If English law is largely case law, then the Bill of Rights made provision for the legal existence of a standing army.

The significance of the resounding assertion contained in the Bill of Rights lies less in its outward appearance and more in its implications. First, it accepts rather than rejects the idea that there will be a standing army in time of peace. This was a major concession. Secondly, the Bill of Rights did not repeal the Militia Acts of the Restoration. These had vouchsafed 'the sole command of the militia and all forces by land and sea' to the king, both in war and in peace. Thus the king's executive power over the armed forces remained untrammelled by the Bill of Rights.[10] It therefore followed that, although parliament had to consent to the continued existence of the army, it retained no say over how that army would be employed.

Very similar points can be made about the other major plank of the 1689 settlement, the Mutiny Act. The Mutiny Act permitted the government to use military law in the punishment of its soldiers. By 1713 it had also become the bill by which parliament authorized the size of the army. In 1881 the revision and consolidation of military law resulted in its replacement by the Army Act. The Army Act, like the Mutiny Act before it, was subject to annual renewal. In 1955 the disciplining of the army and the army's actual size were once again separated, and the Army Act became a quinquennial exercise, although still subject to annual continuation orders. In 1981 it was replaced by the Armed Forces Act, which also runs for a five-year period but with annual continuations.[11]

It was the Mutiny Act which gave rise to the idea that the army existed only on annual sufferance—that, in legal terms, the British army has never had any guarantee of its permanence. The passage of the Act, although latterly a formality, in the eighteenth century acquired the status of ritual, venerated by lawyers and constitutionalists. Sir John Fortescue expressed an irritation which was totally justified: he castigated the Mutiny Act and its successor as 'a ridiculous piece of pedantry'.[12] He was right not just on the grounds of common sense but also as a historian. The Act had acquired accretions that distorted its origins and inflated its significance. It was never conceived as part of the 1689 settlement. Its passage was entirely coincidental.

In March 1689 men of the 1st Royal Regiment of Foot (or Royal Scots), then at Ipswich, declared that they would soldier no more and set off northwards for home. England was in confusion; William was worried about the threat from France; if one regiment mutinied, others might follow. What the Royal Scots had done was punishable in civil law, but that was a slow process at a time when speed and decisiveness were of the essence. Parliament responded with emergency legislation to allow courts martial, as opposed to civil courts, to punish in cases of mutiny, sedition, or desertion. The initial life of the Mutiny Act was six months. The short period reflected the fact that it was a response to a specific crisis; it was not part of a more ambitious plan to oblige the army to be dependent on the Act's annual renewal by parliament. Indeed, by separating military law from civil law it helped distance the army from the civilian population. Thus its effect was actually in part to undermine the arguments of those opposed to standing armies.

The army had existed without a Mutiny Act before 1689, and it was to do so again between 1698 and 1701. No Mutiny Act did not mean no army.[13]

Therefore the 1689 settlement was an extraordinarily positive step forward in the establishment of the British army. It endorsed its existence in peacetime, and it left the king with a remarkably free hand as to how he used it. It took the sting out of the criticisms of the previous sixty years. Parliamentary attacks on the army's existence were not over, but increasingly they represented form rather than substance. In many ways, therefore, the army was freer than it had been hitherto.

However, Clode's emphasis was not on freedom but on controls. Because his ambition was to write a constitutional history of the army, and because that delivered him—effectively bound hand and foot—into the arms of the Whig historians, he saw 1689 as the moment when constitutional controls were established. This created difficulties. Even Clode had to admit[14] that in practice parliament possessed remarkably few constitutional controls over the army. The Mutiny Act became the basis on which it voted the number of men in the army, but it retained no discretion over how the crown or the ministers of the crown used those men. They could send them all abroad, so leaving the home base exposed to invasion; they could concentrate them all in Britain, so posing a latent threat to civic freedoms. In neither case had parliament reserved to itself the right to intervene.

However, we should not conclude that, just because parliament had only limited constitutional controls, it was therefore without checks on the army. It had controls but they were not constitutional, and they were not the fruit of the 1689 settlement; they were financial. Supply was the means by which parliament managed Charles II's army; supply—too much of it—was why it lost control of James II's army.

Therefore, parliament's major confrontation with William III over the army followed not from the Glorious Revolution but from the peace of Ryswick in 1697. Parliament, angered by William's use of English resources to wage war on the Continent, saw the opportunity to make reductions. William, conscious as a Dutchman of the needs of the European balance of power, regarded the peace as no more than temporary. He wanted to keep the army intact. But by 1697 it was already the practice to state the exact size of the army before supply was voted. William's only way out of his dilemma was subterfuge. He hid troops in other establishments, where they burdened other exchequers, and where their costs were not subject to the scrutiny of parliament.[15] Ireland was his immediate bolt-hole. For later generations—after the union of the two parliaments in 1800—India would take Ireland's place.

The practicalities of money, not the formalities of constitutionalism, therefore drove parliament in its pursuit of control. At the same time as the English parliament was debating with the king the possible savings to be extracted from the peace of Ryswick, the Scots began to debate the need for a Scottish militia. Their case was economic, not constitutional. Historians have tended to emphasize the contrast between these preoccupations and those of their contemporaries south of the border. However, there may actually be a revealing similarity. To be sure, Scotland at the beginning of the eighteenth century was more backward than England. But in both countries a prime objective was the harmonization of military service with economic growth. Andrew Fletcher of Saltoun argued in 1698 that Scotland's martial tradition took men away from more productive pursuits, and he therefore favoured the militia over a standing army. Half a century later Adam Smith put the opposite argument in *The Wealth of Nations*. He favoured functional specialization, and so preferred a standing army to a militia. He was dismissive of the political dangers of a standing army 'where the sovereign is himself the general, and the principal nobility and gentry of the country the chief officers of the army'. His phraseology thus anticipates both Clausewitz and Janowitz; his preoccupation was—like Fletcher's—entirely economic. Scotland did not get a militia in the eighteenth century, but nor in many senses did England. If the constitutional danger of the standing army was such a pressing threat south of the border, why did the Whig gentry allow the militia to wither? Not until 1757 was it revived, and then in response to the external threat from France, not to the domestic danger of an over-mighty monarch. The English, like the Scots, saw economic growth as more important than the self-important rhetoric of constitutionalism.[16]

Parliament's aim in the eighteenth century was, therefore, not the establishment of constitutional controls but devices for more cost-effective management. Even Clode recognized this point. He attached enormous significance to an Act of 1783, carried by the paymaster-general of the forces, Edmund Burke, which obliged the crown to bring regimental allowances before parliament as part of the army estimates. The effect of the Act was to remove the prime responsibility for actual expenditure from the officers of the regiment to the secretary at war.[17] What it overcame was the point which William's use of Ireland had highlighted. Parliament might huff and puff until it voted supply, but the 1689 settlement had given it no powers to determine how the money was spent thereafter. Burke's Act enhanced parliament's powers of audit.

More effective scrutiny depended on parliament possessing an officer who could represent its interests and who could act as a bridge between itself and the army. By 1783—and until 1855—that officer was the secretary at war. But the status of the secretary at war was not established in 1689. It evolved during the course of the eighteenth century, and it was the product of practice rather than of constitutional theory.

In origin the secretary at war was not a political officer at all. Certainly he was normally a civilian rather than a soldier, but he was military secretary to the king rather than to parliament. His responsibilities to the king in the latter's capacity as commander-in-chief included piloting the Mutiny Act and the army estimates through the House of Commons. Thus his tasks as a royal servant also gave him parliamentary functions. Under Queen Anne, the effective commander-in-chief, Marlborough, was abroad for much of the time, and therefore the secretary at war combined with his management of parliamentary and financial issues the exercise of royal patronage and the enforcement of discipline. In these circumstances, the administration of the army did not rest on dual control, on the notion of checks and balances favoured by Clode, but on fusion.[18]

The point about fusion remained true even when Marlborough came back from his campaigns. The office of secretary at war did not normally carry cabinet rank, and—largely as a result—it was often bestowed on a political lightweight. The effect of having an active commander-in-chief at home was to downgrade the power of the secretary at war yet further. If the notion of professionalism is to have any meaning in the context of the early eighteenth century, then Marlborough was a professional. But he was also a political figure. He had played a key role in the plot to oust James in favour of William. Thereafter his behaviour was certainly politicized in both a narrow and a broad sense. His survival depended on his nose for political intrigue, and he used the patronage in his gift to support and develop his own career. But he also had an understanding of strategy which escaped the constraints of tactics and operations, and included the formulation of war aims. Logically it led to the development of an independent foreign policy. Thus 'fusion' is a word as appropriate to the nature of his power as commander-in-chief as it is to that of the secretary at war.

For much of the eighteenth century the army had no effective commander-in-chief, and this was particularly the case in periods of peace. The secretary at war was the fixture, and it was therefore his power that waxed. When there was no commander-in-chief, the heads of the subordinate military departments—the adjutant-general (responsible for returns and discipline) and the quartermaster-general (responsible for quartering, encamping, and moving troops)—answered to the secretary at war. Thus effective command and administration were united in one authority: dual control was a fiction.

The most senior soldier was frequently the master-general of the ordnance. This was a more ancient and arguably more prestigious office than that of commander-in-chief. However, again the theme must be fusion. The Board of Ordnance, over which the master-general presided, was at once both civil and military. On the one hand, it was a department of disbursement and supply; it equipped both the army and navy. On the other, it was also a department with responsibilities for command; both the Royal Artillery and the Royal Engineers were subordinated to it. Furthermore, although the master-general was a soldier, his appointment was political. He normally sat in the cabinet, to which—as a soldier—he gave advice on matters of professional concern. He left office when the government fell.

The post of commander-in-chief did not achieve continuity until

1793. The outbreak of war and its duration for over twenty years gave permanence to arrangements that had previously been temporary. The expansion of the armed forces created an enormous field of patronage. In peacetime the danger of having promotion and preferment in political hands was minimized by the operation of purchase. But war not only expanded the scale and scope of promotion, it also rendered purchase less significant in its operation as vacancies arose more frequently through death. Therefore the most important political consequence of the appointment of a commander-in-chief was that patronage in the army was removed from the hands of a political appointee and entrusted to an officer of the crown. In 1795 the king appointed his son, the duke of York, as commander-in-chief. The king's power in relation to the army was thus enhanced, while that of parliament was correspondingly diminished. Dual control had at last begun to acquire genuine substance; its consequence was friction.

In 1809 the duke of York was forced—albeit temporarily—to resign his office. Significantly the scandal that caused his fall involved his powers of patronage: his mistress, Mary Anne Clarke, claimed that she had run a successful business receiving payments from hopeful officers in exchange for suggesting their promotion to the duke. General Sir David Dundas took over the reins at the commander-in-chief's office at the Horse Guards.

Dundas took his stand on the practices of a century previously, arguing that the secretary at war was the subordinate of the commander-in-chief. The law, *pace* the Bill of Rights, was confused, but the warrant appointing the secretary at war did indeed require him to 'follow such orders as he shall from time to time receive from His Majesty, or from the General of His Forces, according to the discipline of war'.[19] The fact that an officer, Lieutenant-General Sir James Pulteney, was secretary at war when Dundas arrived at the Horse Guards no doubt helped confirm Dundas in the justice of his position. But in October Pulteney was succeeded by a politician of different mettle, Henry Temple, Lord Palmerston.

Palmerston remained secretary at war for all but twenty years— in which the commander-in-chief disputed the secretary at war's pecuniary powers, and the secretary at war resisted any subordination to the extra-parliamentary authority of the commander-in-chief. Palmerston was convinced that the situation was a new one, dating 'only from the command of the Duke of York': he made no reference to 1689. The duke

was a strong man; son of one King & brother of the other; heir presumptive; a political leader; he commanded too in time of war, when all men's minds took a military turn, & were accustomed to defer to military authority; he was always at the head of the army (except during a short interval) and took advantage of every opportunity to push on his encroachments; he had to do with a frequent succession of Secretaries of [*sic*] War, each of whom had to learn his duties & his powers, and many of whom had their pickets driven in, before they had got well into their saddle.[20]

In 1810 Palmerston seems to have sought a showdown, but in May 1811 Dundas stepped aside to allow the duke of York to return to the Horse Guards. The duke's brother, now prince regent, was keen to buttress royal power, and, although Spencer Perceval, the prime minister, was anxious not to compromise the authority of parliament, he recognized that it was not appropriate to expose the divisions between military and civilian control of the army in the middle of a war. The upshot was a warrant, signed on 29 May 1812, which addressed the issue but did not resolve it. It stated that the secretary at war was a financial officer responsible to parliament, while the commander-in-chief was entrusted with discipline and was answerable to the crown. The warrant went on to encourage the two officials to talk to each other—a request that over the rest of the nineteenth century was met with singular infrequency.[21]

Herein was the first clear statement as to what dual control was, if not how it was meant to work. The secretary at war should be a civilian, and the commander-in-chief should not be a politician. The commander-in-chief was responsible to the cabinet because it was composed of ministers of the crown; he was not responsible to parliament because that would have been an encroachment on the royal prerogative. After Waterloo this division of responsibility acquired the patina of age and of constitutional precedent. As Palmerston pointed out, the duke of York had created 'a system which you may be sure was not relaxed by the Duke of Wellington when he succeeded to the office'.[22] In 1837 Wellington duly told the prime minister, Lord Melbourne, that 'I have always understood that it was a principle of the government of this country, that he who exercised the military command over the Army should have nothing to say to its payment—its movement—its equipment—or even the quartering thereof excepting under the sanction of a civil officer who has himself subordinates in the hierarchy of civil office, and could not take the King's pleasure excepting upon matters of account'.[23]

The great merit of this division of responsibility, Wellington emphasized, was that political influences were kept clear from matters of military patronage and promotion. Therefore, he concluded, 'I have always been of opinion that the Commander-in-Chief ought not to be a member of the Cabinet; . . . the military resource for the Government is the Master General of the Ordnance.'[24]

In reality past practice had been very different from the principles which Wellington had so dogmatically enunciated. The interpenetration of soldiers and politicians in the eighteenth century was not confined to the junior ranks. Commanders-in-chief before 1795 had not felt themselves disqualified from political office or political activity by virtue of their military appointment.

After Marlborough's demise in 1722 there was no 'captain-general' of the army until the appointment of the earl of Stair as commander-in-chief in 1744. The duke of Cumberland was appointed to the office in 1745. Between 1754 and his resignation in 1757 Cumberland and Henry Fox, the secretary at war, organized a group of MPs in the House of Commons, most of them officers. Both Cumberland's successors as commander-in-chief, first Lord Ligonier and then the marquess of Granby, were simultaneously also MPs and masters-general of the ordnance. Granby emphasized his political loyalties by resigning in 1770, when Chatham went into opposition. In 1782 Henry Seymour Conway, mentioned in the previous chapter as the last soldier to be deprived of a military appointment by the crown for his behaviour in the House of Commons, became commander-in-chief in England. A member of Shelburne's government, he withdrew from the cabinet on the formation of the Fox–North coalition, and in December 1783 resigned from his office as he believed Pitt's assumption of office was unconstitutional. At the same time, Sir John Burgoyne, appointed by Rockingham to the more junior post of commander-in-chief in Ireland in March 1782, began to show an awareness of some of the constitutional points adumbrated by Wellington. When Shelburne formed the government, Burgoyne responded to advice not to give up his appointment on the grounds that it was not political: indeed Lord Portland suggested to him that he could both hold on to his office and oppose the government. But this was having it both ways. In January 1784 Burgoyne resigned as he had felt his office was now politicized. The post of commander-in-chief in England remained vacant between Conway's departure in 1782 and Lord Amherst's appointment in 1793. Amherst had been commander-in-chief before, between 1778 and 1782, but he had not then sat in parliament. However, in 1793 he was simultaneously commander-in-chief and a member of the cabinet.[25]

The notion that the commander-in-chief was not to hold political office was therefore of no greater antiquity than the tenure of that office by the duke of York. Moreover, even he had found the division between political affairs and military too crude to be helpful. His answer to the Mary Anne Clarke affair and to the 1812 warrant had been to get his own representative into the House of Commons in order to have a spokesman to explain and justify his administration of the army. Thus the military secretary, the commander-in-chief's right-hand man, who in particular advised on promotions and appointments, frequently sat in the Commons—Lord Fitzroy Somerset (the future Lord Raglan) was MP for Truro from 1818 to 1820, and again from 1826 to 1829, and Sir Herbert Taylor was MP for Windsor between 1820 and 1823. Lord Liverpool, the prime minister, expressed his disapproval of the arrangement in 1820, and told Taylor he was not to interfere in discussions on the army estimates.[26]

However, the real exemplar of this fusion of military and political activity was Wellington himself. His practice was at direct variance with his precepts. Wellington liked to insist that he was primarily a soldier. But such posturing made his friend Mrs Arbuthnot very cross. 'It was ridiculous nonsense', she told him, 'for him to stand up and tell he was *no politician* . . . that he had taken a most active part in both our home and foreign politics; that a body in the State looked to him as their protection against the political economists, and that he must be quite aware that they did not do so on account of his military talents.'[27]

Wellington's political pedigree, if inevitably not so distinguished as his military, had developed in tandem with it. He sat in the Irish House of Commons between 1790 and 1795, and in 1805 he had entered the British parliament—principally in order to defend his brother Lord Mornington. In the process he became Irish secretary for two years. The Peninsular war interrupted his parliamentary career but—like Marlborough before him—senior command honed rather than rusted his political skills. The need to ensure consistent governmental backing at home, the negotiations with his Spanish and Portuguese allies, and then the settlement of Europe after Napoleon's defeat—all of these were tasks that were more political than strictly military. In 1818 he was appointed master-general of the ordnance, a post he held for well nigh a decade, and which still combined military and political responsibilities, and carried with it membership of the cabinet.

In 1827 the duke of York died, and Wellington succeeded him as commander-in-chief. According to the principles which he declared a decade later, he should now have resigned from the cabinet. He did not do so. Wellington's worry was the succession to the premiership on the departure of Lord Liverpool. The Ultra Tories told the king that Wellington himself could be prime minister while still remaining commander-in-chief. Wellington himself probably did not covet Liverpool's job, and briskly said (at least on this occasion) that the royal prerogative would prevent him holding both appointments. What kept him in the cabinet was his anxiety for its unity and his desire to block the path of Canning.

Canning's appointment as prime minister shattered that unity. Wellington resigned—not only from the cabinet but also from his post as commander-in-chief. He continued to protest that differences of political opinion should not weigh on the incumbent at the Horse Guards, but then said he himself could not stay in the job if it required him to have close dealings with Canning. Confirmation that the duke held to one rule for the generality and another for himself was not long in coming. In August 1827 Canning died. The king then invited Wellington to resume the post of commander-in-chief, and Wellington accepted.

It was therefore totally in line with his previous behaviour that Wellington should have tried to remain commander-in-chief when he became prime minister in 1828. Indeed he said he wished that he had made his retention of the Horse Guards a condition of his acceptance of the premiership, and declared that if he had been given the option he would have preferred the military office over the political. On this occasion, however, the cabinet would have no truck with his holding both at the same time. Wellington gave in but with a bad grace. His response was to bring two distinguished soldiers into his government, Sir Henry Hardinge as secretary at war and Sir George Murray (who had been quartermaster-general in the Peninsula) as president of the Board of Trade. The military ethos of the duke's prime ministership was confirmed by his decision to resolve his differences with Lord Winchilsea in a duel, in which Hardinge was his second. Such action was more characteristic of soldierly codes of honour than of parliamentary behaviour.[28]

The military complexion of Wellington's cabinet was of course in line with the reputation of its chief. This was a point which Canning had appreciated in 1827. The formal renunciation of the Horse Guards would not actually be enough to distance Wellington from the command of the army. The very fact of his being prime minister would—because he was also Britain's most distinguished general since Marlborough—inevitably unite, as Canning told Wellington, 'the whole power of the state, civil and military, in the same hands'.[29] Wellington was honest—or vain—enough to acknowledge the truth of Canning's observations when he did become prime minister. He confessed to his cabinet that 'all he was in the world was owing to the Army', and that therefore his 'connection with it never could cease be his situation what it might, & that he must always in truth & in fact & morally be the C.-in-C. of the Army of which he was the Senior Officer'.[30] Wellington's involvement in politics was inevitably in opposition to the nostrums of dual control, and could not be gainsaid by a preoccupation with technicalities.

When Sir Robert Peel formed his second government in 1841 he invited Wellington to be the leader of the House of Lords and a member of the cabinet without office. In the following year Lord Hill, who had succeeded Wellington at the Horse Guards in 1828, was forced to give up office through ill-health. Wellington was very anxious to resume the post of commander-in-chief. He dutifully went through the motions of saying he should resign from the cabinet in order to do so. However, Peel insisted that he stay in the government and that he continue to speak on its behalf in the upper house. Thus once again Wellington technically as well as practically combined military and political office. He did not resign from the Horse Guards in December 1845 when Peel resigned, but any suggestion of political neutrality contained in this move was quashed in the following year. When the Whigs rejected Peel's challenge that they should repeal the Corn Laws, and passed the 'poisoned chalice' back to Peel, Wellington saw it as his duty to support the Conservative leader in the Lords by steering the measure through the upper house. Again, technicalities were used to mask the substance. Wellington argued that he was supporting the government of the queen and not the ministry of the day.[31]

The smashing of the Conservatives on the rock of the Corn Laws did finally close Wellington's political career. But it did not end his involvement with the army. Wellington was a ministerialist; he saw his task as the maintenance of the queen's government. If the Conservatives could not form a ministry, then Lord John Russell's Whigs were the next best thing. The sophistry that permitted him to support Peel in repealing the Corn Laws also enabled him to continue as commander-in-chief under Russell. In this respect December 1845 had been a dress rehearsal. He remained commander-in-chief until his death in 1852.

The inconsistency between Wellington's public declarations on the subject of dual control and the pursuit of his own career is obvious. His behaviour as commander-in-chief was consonant with, if more exaggerated than, that of his eighteenth-century predecessors. But he lived in an era when the pronouncements of Mackintosh and Macaulay hallowed the arrangements of 1689 with the aura of constitutional precedent. They were Whigs, and they put forward a Whig interpretation of the past. Wellington was a Tory, and yet he elected to defend a Whig settlement, so giving the principle of dual control a clarity and definition that it had never previously possessed.

Wellington's pronouncements were motivated by three interlocking factors. First, Wellington recognized, as Palmerston did, that the elevation of the office of commander-in-chief by the duke of York had made the incumbent at the Horse Guards (in Palmerston's words) 'the head of a great civil as well as military department without any direct contact with Parliament so as to render him practically responsible'.[32] Secondly, therefore, the idea of dual control was a way of preventing parliament taking over and centralizing the administration of the army. Twice the Whigs attempted this—in 1837 and again in 1849—and twice Wellington was in the van of the opposition. He was driven not just by his fear of the consequences for the army estimates, but also by his ministerialism. His duty as he saw it was to sustain the king's or queen's government. He therefore cleaved to the 1689 settlement in order to protect the royal prerogative. His concern was not to defend the power of parliament in relation to the crown, but the power of the crown in relation to parliament.[33]

A clear illustration of his anxieties concerning the crown was his effort to ensure his own succession. He wanted Prince Albert to follow him at the Horse Guards. He argued that a commander-in-chief of royal blood was required in order to preserve the queen's prerogative and 'to counteract the growth of Democratic Power'. Wellington's enthusiasm for the notion of dual control was thus laid bare: his underlying concern was 'that the command of the army should remain in the hands of the Sovereign, and not fall into those of the House of Commons'.[34]

When Wellington put this proposal to the prince consort in 1850, Albert resisted it. He could see that the duke's plan was likely to backfire, that it could bring opprobrium rather than approbation down on the head of the monarch. None the less form and substance should—once again—be distinguished. Albert acknowledged that, in view of the monarch's sex, he might reasonably take an active role in military affairs. When Wellington died, the prince consort's involvement in the reform of the army was deep and intense, sufficient to excite press speculation and to require statements in parliament.[35]

The commander-in-chief actually appointed in 1852 was Hardinge. A Peelite, twice secretary at war (1828–30 and 1841–4), governor-general of India (1844–8), and currently master-general of the ordnance, Hardinge no more than Wellington saw the achievement of high political office as a block to his incumbency of the most senior post in the army. Like Albert, he recognized that the army was in need of vigorous reform, but, like Wellington, he did not see this as embracing its unification under civilian control. In 1837 and again in 1849 Hardinge had supported his former chief in his efforts to check the consolidation of army administration.

But with the outbreak of the Crimean war in 1854 his resistance collapsed. In June, to silence further Whig criticism, Lord John Russell set in motion the steps which would make the secretary of state for war and the colonies—a post first called into existence by Pitt in 1794—a minister of war. The responsibilities for war and the colonies were divided, and the secretary of state—the duke of Newcastle—opted to take charge of war. In December the commissariat—which in the field had responsibility for the government chest as well as for supply—was transferred from the Treasury to the secretary of war. Hostilities also exposed the frailties of the Board of Ordnance. Hardinge's successor as master-general, Lord Raglan, went off to command in the Crimea without renouncing his office: the effect was to leave the Board of Ordnance in disarray. The members of the Board put themselves in Newcastle's hands even before their formal transfer in 1855. This accretion of military responsibilities by the secretary of state made the secretary at war not only subordinate but also increasingly redundant: the two offices were amalgamated in February 1855, and that of secretary at war was formally abolished in 1863.

Ostensibly these reforms served to strengthen the civilian and political control of the army. But the corollary of the consolidation of the civil administration was a sharpening in definition of the military administration. The status and the responsibilities of the commander-in-chief were clarified; the ambiguity created by the fusionist practices of the eighteenth century and by the career patterns of Wellington and Hardinge was largely resolved.

What above all ensured that the Horse Guards were not absorbed as the other departments were was the reassertion of the royal prerogative. In 1834 Hardinge had insisted that 'it was of the utmost importance that the army should look up to no authority but that of the King'.[36] In 1854 he successfully defended that principle. And in 1856, when ill-health forced his resignation, it found practical embodiment in the appointment of a royal prince as his successor.

When the queen's cousin, the duke of Cambridge, became commander-in-chief, he was seen—in contradistinction to his later reputation—as young, professional, and reform-minded. The duke refused to regard himself as subordinate to the secretary of war, even if the latter's letters patent suggested he should be and even if it was argued that the minister was the queen's minister as well as the member of a government accountable to parliament. In 1860, the select committee on military organization recommended that, in the event of conflict between the commander-in-chief and the secretary of state for war, the final decision should rest with the latter. The duke rejected this suggestion, insisting that matters of command and discipline were dependent on the royal prerogative, and therefore lay between him and the queen.

The proposals of 1860 were revisited and reaffirmed in 1869, by a committee chaired by the under-secretary of state for war, Lord Northbrook. Thanks to Edward Cardwell, the secretary of state, they were given effect in the War Office act of the following year. The commander-in-chief was now designated the principal military adviser to the secretary of state, and his department was one of three directly subordinate to the minister. The duke's office was moved from the Horse Guards to the War Department. But, if the physical environment changed, the duke's attitudes did not. He remained wedded to the principle of military independence, and he contrived to be at odds with all three of the significant secretaries of state between 1870 and his retirement in 1895—Cardwell, Hugh Childers, and Edward Stanhope.

In 1888, a year into the latter's tenure of office, the government responded to the army's own anxiety about the military estimates by setting up a commission on the administration of the two services under the chairmanship of Lord Hartington. In 1890 the Hartington commission recommended that the post of commander-in-chief be abolished, and that the secretary of war should in future be advised by a joint military–civil War Office council. The soldiers to serve on the latter would include the adjutant-general, the quarter-master-general, the inspector-general of fortifications, the director of artillery, and the chief of staff. The last-named was to be the head of another innovation—a general staff responsible for considering the issues of imperial defence and for concerting the plans of the army and the navy.[37]

The duke's outraged response to these proposals was matched by that of the monarch. Over a decade previously, in 1874, the secretary of state for war, Gathorne Hardy, had been struck by the queen's insistence that 'her name [be] conspicuously printed in the Army list, on the first page, with a page to itself and some note implying that she is the head of the army: with her A.D.C.'s names given in the next page, before any mention of the Sec. of State or C.-in-C.' Hardy thought there was 'some craze in all this', but Lord Derby was more prescient: 'there is design also, and the question may yet give trouble.'[38]

The queen thwarted the duke of Cambridge's wish to retire, anxious that he should keep the commander-in-chief's seat warm until Prince Arthur, duke of Connaught, was of age to take over, so 'that it may not go out of the family'.[39] The Hartington commission's proposals threatened to wreck these schemes. She promptly reminded her ministers that 'one of the greatest prerogatives of the Sovereign is the *direct communication* with an immovable and non-political officer of high rank, *about the Army*'. So powerful was her opposition that she secured the postponement of any change until 1895, the year in which the duke of Cambridge retired. Furthermore, she ensured the survival of the post of commander-in-chief.

The government was of course perfectly well aware of this bond between the army and the crown, and did as much as it could to break it. In the year when Hartington's commission reported, Lord Roberts, the commander-in-chief in India, was due to come home to be adjutant-general. But his potential successor in India was the duke of Connaught, and Stanhope feared that the post of commander-in-chief in the subcontinent might prove a stepping-stone to that at home. The secretary of war concluded that Roberts's tenure should be extended rather than elevate another royal prince. Over the next five years the duke of Connaught found his military career treated as a shuttlecock between the two factions, the queen trying to manœuvre him into appointments of influence and the politicians endeavouring (with greater success) to block him.[40] In 1895, when the duke of Cambridge retired and some at least of Hartington's recommendations began to be effected, Stanhope's successor, Lord Lansdowne, offered the now-vacant post of commander-in-chief to Garnet Wolseley. He did not consult the queen: Victoria protested, not only over the issue of principle but also over the individual chosen.[41] Wolseley soon discovered that under the new arrangements the grand title of his job was not matched by the authority which it conveyed. In December 1899, when Lord Roberts was appointed to command in South Africa, neither the queen nor Wolseley was asked for an opinion.[42]

However, the link between the army and the crown survived the 1895 changes. Wolseley deflected responsibility for the initial defeats in South Africa onto Lansdowne. His public lobbying made clear that the establishment of the War Office council had not removed the adversarial legacy of dual control—that civilian and soldier still pitted themselves against each other. In October 1901, his protests were rewarded with some success: in a policy designed to encourage decentralization, authority was conceded to soldiers at all levels of command, and the three most important military departments were made answerable to the commander-in-chief. Furthermore, in 1902 the government established a royal commission under the chairmanship of Lord Elgin. Its brief was to consider the South African war, and in particular to take account of the accusations levelled by Wolseley against the secretary of war.

One of the commissioners was Reginald Brett, Lord Esher. A courtier, he had, after 1895, taken the place of the duke of Cambridge as an informal conduit between the monarch and the army. As the soldiers presented their evidence to the Elgin commission, Esher conveyed its essence back to the king. The generals blamed the civil authorities, and Esher spared no efforts in reinforcing this interpretation and in stimulating the king's interest in military matters. This was not as difficult a task as Edward VII's lackadaisical reputation might suggest. His interest in army reform had been stirred by Hartington in the 1880s, and the effects of the South African war confirmed him as a strong advocate of change: in this respect he took after his father rather than his mother. But royal intervention, although it favoured reform, was politically even more suspect than it had been fifty years previously. Thus the king needed the army just as much as the army needed the king: both had a mutual interest in curbing the civil authorities.[43]

Esher acted as the broker in this relationship. When the adjutant-general, Sir Evelyn Wood, appeared before the commission, he reminded Edward of the 'very superior intellectual equipment of this officer, and the singular precision of his ideas on all military subjects'. He concluded his report by saying, 'the gist of his evidence, like that of every other officer hitherto examined, was that for all the shortcomings at the W.O. it is Your Majesty's Civil Servants, and not the military Officers, who are to blame'.[44]

On 4 December 1902, the commission saw Lord Roberts, who had succeeded Wolseley as commander-in-chief in 1901. Esher stressed that the impression left by Roberts 'cannot fail to have been that the Commander-in-Chief of Your Majesty's Forces, although nominally in a position of great responsibility, is so hampered by the organization of the War Office, and by the Parliamentary customs which hem that officer within an impassable fence, that his responsibility is an illusion'. Esher's conclusion was that 'this . . . is the gravest question connected with the organization of Your Majesty's Army'.[45]

The commission had taken evidence from Wolseley himself the previous week. The field marshal was adamant that the responsibility for the disasters of 1899 lay with the cabinet because it had no machinery which enabled it to hear directly the views of the commander-in-chief: all was filtered through the secretary of war. His solution was twofold. The secretary of state for war should be a soldier, and the king himself should be commander-in-chief.[46]

Esher was quick to dismiss Wolseley's ideas on constitutional grounds. But in reality his own thinking developed in directions that were not unrelated. Initially, he had felt that the commander-in-chief should be freed of administrative duties so that he could concentrate on executive and operational matters, readying a general staff for war rather than worrying about the financial constraints of peace. But then, perhaps in response to Wolseley's views, he saw in the post scope for reasserting the royal prerogative. Edward's own inclination was, in this respect at least, to follow the course favoured by his mother, and appoint the duke of Connaught as commander-in-chief in succession to Roberts. Esher, on the other hand, now concluded that the powers which had been wielded by the duke of Cambridge were not rooted in antiquity but were a temporary phenomenon, the consequence of Britain being ruled by a queen rather than by a king. In May 1903 he reminded his royal master 'that Your Majesty is the natural Head of the Army all over the world'.[47]

The corollary of the king becoming commander-in-chief was of course the abolition of the post as it had developed after 1895. It would no longer be the billet of Britain's senior soldier. Such thinking was compatible with the conclusions of the Elgin committee. The latter had not, for all the volubility of the soldiers, exculpated the commander-in-chief. It had acknowledged that his office was inadequate for the tasks of a general staff, one of the recommendations of the Hartington commission which had still not been implemented. However, much to Esher's fury, the Elgin committee had not gone on to propose any substantive solutions.

In September 1903 H. O. Arnold-Forster, a politician well versed in military affairs, was appointed secretary of state for war. Arnold-Forster saw the reorganization of the army's recruiting as more important than the restructuring of the War Office. This is not to say he did not have views on the latter. He favoured a revivified form of divided authority: the professionals should be given their head as advisers of the government and should be provided with a general staff to enable them to formulate that advice. Arnold-Forster also saw it as important to keep the king at a distance: any professional communication regarding military matters should be channelled through the secretary of state alone.[48]

Arnold-Forster's ideas were unacceptable to Esher. He had now concluded that the genuine integration of military men and civilian advisers should be the goal. This was a direction in which the Committee of Imperial Defence, set up in 1902 through the agency of the prime minister, Arthur Balfour, was already pointing. Esher had refused the War Office before it was offered to Arnold-Forster. In its place he secured the chairmanship of a committee to review the army's administration, from which Arnold-Forster, despite the fact that he was the man responsible for military affairs to both cabinet and parliament, was excluded. Esher used his chairmanship to sideline the secretary of war in particular and the cabinet in general. Instead he consulted the king, seizing the opportunity to reassert the royal prerogative.

In 1904 the Esher committee reported. It proposed that both the War Office council and the commander-in-chief should go. In their place, an Army Council—inspired by the fusion of civil and naval officers on the Admiralty Board—should unite civilians and soldiers together in one body. Although the summary removal of Lord Roberts from his post as commander-in-chief gave the impression that the independent voice of the army was once again under attack, the reality was the opposite. The Army Council consisted of seven members, of whom four were professionals: the soldiers therefore had the majority. The finance department of the War Office was castigated and its size and powers were reduced. And the creation of a general staff gave professional advice real clout for the first time.

More remarkable in the context of relations between the army and the crown was the establishment of a second new major military post, that of inspector-general of the forces. Roberts hoped that this might be a role for him, but was told that if he took it on it would be hard to resist the implication that it was a reincarnation of the office of commander-in-chief. Roberts—not unreasonably— suspected a plot hatched by Esher to make the king the effective head of the army.[49] These thoughts found confirmation when the duke of Connaught was appointed inspector-general. It was not just that the duke was a royal prince and a field marshal; Esher also contrived, in the face of fierce cabinet opposition, to get the post of inspector-general tenable concurrently with that of president of the appointments board.[50]

Esher hoped to bring at least the senior levels of military patronage under the active control of the crown. Engaged on the task of editing Queen Victoria's correspondence in 1905, he confessed himself struck by 'the remissness of the King's present advisers in not keeping His Majesty informed, and in not sending their written reasons or seeking the King's authority BEFORE action is taken'. What this meant in relation to the army was that 'there is always to be developed as time goes on the authority of the King as Commander-in-Chief. I mean in all personal questions. The King should adhere tenaciously to his right to veto any appointment.

Gradually it will become clear to everyone that under a *King* a C-in-C was an anomaly.'[51]

Thus the link between the army and the crown survived the move from dual control to integrated control, albeit in an attenuated form. Esher's more extreme hopes were not fulfilled, and Arnold-Forster immediately set about the limitation of the duke of Connaught's powers. But the army itself persistently brought the monarchy back into the frame, using it as a device to fend off governmental intervention. And Edward VII and then his son George V colluded. Outwardly they protested their desire to preserve their constitutional position by remaining above such wrangles. In reality they never divorced themselves from the notion, which the army assiduously cultivated, that the army was 'theirs'.

In 1914 those officers who refused to resign over Ulster during the Curragh incident defended their position not on the grounds that they should be the loyal instruments of a democratically elected government but on the principle that they should obey the king. The officer who did most to contain the effects of the incident was Sir Charles Fergusson, the major-general commanding the 5th Division in Dublin. Privately, Fergusson confessed that his sympathies were with the 'mutineers', but he saw his public duty in different terms because he laboured under the misapprehension that the orders to proceed north, and the 'ultimatum' which obliged all officers but those domiciled in Ulster to do so, originated with the king. The man responsible for creating this impression was the general officer commanding in Ireland, Sir Arthur Paget, with his public insistence that he did not take orders from politicians, only from the sovereign.[52]

What motivated the army in 1914 was professional self-regard. What worried Fergusson was division and disunity within the officer corps. Such priorities allowed senior officers to impute to the king views which he had not expressed, and to do so without seeking corroboration until after the damage had been done.

One of the officers to whom Esher had been particularly anxious to draw royal attention, from as early as 1903, was Douglas Haig.[53] Haig was ADC to Edward VII, who as prince of Wales was friendly with his sister. He was encouraged by the king to report directly to him when appointed inspector-general of cavalry in India. In 1905

Haig came home on leave and stayed at Windsor for Ascot week. While there, he met and decided to marry the queen's lady-in-waiting, Dorothy Vivian. The royal connection survived the death of Edward VII. When the First World War began, Haig, first a corps commander and then an army commander, corresponded in confidence with George V. When Haig's offensive at Loos in September 1915 miscarried, he was quick to blame his superior, the commander-in-chief of the British Expeditionary Force, Sir John French, for his alleged failure to release reserves in time. By the following month Haig had concluded that Sir John French was no longer fit to command; it was first to Esher and then to the king himself that he confided these views. Prompted by Haig, Esher and the king, not the prime minister and the secretary of state for war, orchestrated French's dismissal and his replacement by Haig.[54]

In December 1915 another confidant of the king, Sir William Robertson, was appointed chief of the imperial general staff. In the struggles between the soldiers and politicians which followed, the king remained a continuous, if shadowy, influence. When in 1916 Robertson needed additional support against Lloyd George—to prevent British troops being diverted from the western front to Salonika, to head off a suggestion that Robertson join a military mission to Russia, and, above all, to still criticism concerning the conduct of the battle of the Somme—he turned to the king. Lloyd George duly backed down.[55]

In the new year of 1917 the king's position became more open. Lloyd George, impressed by the new French commander-in-chief, Nivelle, perversely decided that the latter could unlock the western front, and agreed to a French request that the British army be subordinated to French command. As Nivelle was answerable to the government of France but not of Britain, Lloyd George's undertaking had strategic and constitutional implications which went far beyond either the *amour propre* of the British generals or the operational preferences of Haig. Haig wrote immediately to the king. For the king the issue was of course that of the royal prerogative: his army had been subordinated to a foreign power without his approval. To Haig came royal messages of reassurance, and when the king himself saw Haig he confirmed that he would support him 'through thick and thin'.[56] To Lloyd George the monarch showed no such sympathy, and privately the king's circle saw the prime minister as resolved to break up the monarchy and establish a republic.[57]

The effect of Lloyd George's effort to force Haig into subordination or resignation backfired. Haig used his connection with the crown to consolidate his position in 1917, just as he had used it to promote his career before the war and to secure the supreme command in 1915. It was a subject on which George V was understandably sensitive. When, at the war's conclusion, Haig wished to publish his appreciation of the king's continued confidence in his command, the king decided against such revelations. He feared that Lloyd George would interpret the links as 'unconstitutional'.[58]

In the Second World War George VI acted as a support for officers of the general staff who felt able to speak to him frankly and openly about their tribulations. Alan Brooke, as chief of the imperial general staff, found these audiences particularly sustaining.[59] But no British general since Haig has pulled the strings of the royal prerogative so frequently or so effectively. That this has been so is testimony to the effectiveness with which integrated control has replaced dual control. Soldiers and civilians have come to operate through the joint bodies of which both the Committee of Imperial Defence and the Army Council were the precursors. That aspect at least of Esher's vision has found fulfilment. But ultimately the corollary of that success has been to marginalize rather than to elevate the monarchy in military affairs. The army colluded in Esher's vision of the future role of the crown in military administration because it needed a counterweight to the government. When it no longer needed it, the crown became redundant.

This is not to say that during the twentieth century there has been any diminution in the symbolism of the link between the monarchy and the army. However, it is perpetuated through regimental links—through royal colonelcies, through the presentation of colours—rather than through high policy. In 1946 Montgomery, then chief of the imperial general staff, found his plans to cut the Brigade of Guards in the same proportion as the infantry of the line thwarted when the major-general commanding the brigade appealed directly over his head to George VI.[60] And even in 1991 the army, hoping to head off the consequences of post-Cold War reductions, made public the unhappiness of the queen and of other members of the royal family over amalgamations which would affect regiments of which they were colonels.[61]

Thus in 1990 a retired brigadier, Peter Dietz, could write: 'Most army officers in Britain, at least until recently, have believed that they stand in a special relationship with the monarch and that this relationship confers privileges of an extra-constitutional nature.' The caveat in Dietz's opening phrase did not derive from any suggestion that his concluding words reflected a situation that was now seen as inappropriate by either of the two parties involved. Rather, Dietz's doubts were prompted by the possible professional self-interest of the army. As the royal family courted the popular press, and in return found its lives and loves exposed to media speculation, its authority was undermined. Dietz could, in these circumstances, envisage a situation where the army's close association with the crown might prove counter-productive, so forcing the army to reconsider its position.[62]

Dual control survives to the extent that the crown sees the armed services as an arena where its involvement can be at once both more public and more intimate than in any other walk of life. Today its political significance survives only at the margins. However, in its heyday, in the hands of Wellington and of his Victorian successors, it became an agent not of political neutrality but of politicization. The symbiotic relationship between the monarchy and the army served both of them: it allowed the former to renew the royal prerogative and it allowed the latter to resist consolidated civilian control.

In the navy, Sir James Graham's reform of the Admiralty Board in 1832 established the principles of integrated control seventy years before its application to the army. Between then and 1904 dual control acquired in the army an actuality it had never previously possessed. The balances which were held out as the legacy of the 1689 settlement gained system and *gravitas*; they were revered as constitutional precedent. By using Whig arguments, the army was able to head off Whig and Liberal administrative reform. In the process, Whigs themselves were hoodwinked into believing the nostrums of dual control.

Indeed it is a tribute to the success of Wellington and his acolytes that the defence of dual control formed a principal thrust of Charles

Clode's book. For him the centralization of army administration in 1854–5, however incompletely achieved, threatened to unbalance the constitution. Consolidated civilian control was just as dangerous as consolidated military control. Changes in the 1689 settlement, Clode warned, would lead towards 'that state of Anarchy in which, after the destruction of the Sovereign and Parliament, Freedom was lost, and Military Power here—as at times and in other places it has too often—reigned Supreme'.[63]

Clode failed to see that dual control was really tripartite control. It allowed the army to play off the royal prerogative against ministerial authority. Furthermore, in Clode's day, these struggles were no longer primarily a battle about finance. In the eighteenth century the parliamentary rhetoric concerning standing armies was frequently a device, a cloak for efforts to reduce the army estimates. In the nineteenth century, the funding of the army was addressed increasingly on its own terms. Thus the struggles over the command and administration of the army became separate from those of its cost. The former were exactly what they seemed; they were therefore fraught with political consequences.

4

ARMY AND EMPIRE: INDIA AND THE NAPIERS

Since 1815 Britain has engaged in only three wars on the European continent—the Crimean war (if that counts, given its geographical location) and the two world wars. But barely a year has gone by when the British army has not been fighting somewhere, and very often in more than one place. Its campaigns have tended to be, and even continue to be, on the fringes of empire. In 1967 Harold Wilson's Labour government announced Britain's withdrawal from east of Suez. Nonetheless the two major wars in which the army has been engaged since then, Northern Ireland apart, have been in the South Atlantic and in the Persian Gulf. Extra-European warfare has been the army's dominant experience.

The empire was therefore the most consistent and most continuous influence in shaping the army as an institution. It was the principal agent in the army's acquisition of professionalism. A long-service force was needed to garrison the empire, and the empire provided it with the sustained exposure to warfare that could only be intermittently vouchsafed by the continent of Europe. Any analysis of the army's development that focuses on London or on preparations for war on the Continent must be partial and incomplete. And yet that has been the tendency in British military history. The empire gets forgotten.

Moreover, even if two unlikely conditions are fulfilled—that the empire is mentioned and that the army's politics are discussed, then the empire is only adduced as a further factor in explaining the army's apolitical stance. According to this sort of thinking, the colonies kept the army away from the centre, and they kept it busy. They therefore worked against political interventionism rather than promoted it.[1] This interpretation will not do.

The soldier in the colonies was more than a soldier. He was also a civilizer and a settler; his role was not only to conquer, but also to administer the land he had conquered. And in many territories— even once conquered and settled—there remained the possibility that internal rebellion or trouble on the frontier would yank the soldier back from the tasks of settlement to his primary job of war.

The most coherent exposition of these functions was provided not by a British soldier but by a Frenchman. In 1900 Hubert Lyautey published a famous article in the *Revue des deux mondes* on 'the colonial role of the army'. He conjured up an image of the soldier that was admittedly somewhat fanciful, and certainly more enlightened than his own later practice in Morocco: the warrior, he averred, was also a supervisor of works, a teacher, an artisan. He quoted with approval the words of his own mentor, Joseph-Simon Gallieni: 'the best way to arrive at pacification in our new colony is to employ the combined action of force and politics.'[2] Therefore the implication of Lyautey's article was that colonial warfare of itself politicized the soldier.

Most British soldiers for much of the time did not experience open warfare. But neither was the situation completely peaceful. Ultimately, government rested on force of arms rather than on consent. Philip Stigger, in considering the career of one middle-ranking officer, Major L. F. Knollys, has made the point that the primary role of colonial police forces was not, as it was in Britain itself, the prevention of crime and the maintenance of civil law. It was the reinforcement of the army in the event of external attack or domestic unrest. Many officers—like Knollys—left the army for careers in colonial administration or colonial policing. The one reinforced the others. Stigger concluded his consideration of Knollys's career in terms which not only echoed Lyautey but were also reflective of more recent counter-insurgency practices: 'Colonial administration always had political overtones. Colonial political activity in the broadest sense can be evaluated only in relation to the military and the police presence in the territory concerned. Equally the British army's involvement in any colony is incomprehensible without reference to administrative and police activity. The trinity formed by the security services cannot properly be ignored.'[3]

Military responsibilities and civil administration were therefore integrally linked. Knollys was a small cog, indicative of a major phenomenon. Its more obvious and public manifestations occurred higher up the tree. The jobs of commander-in-chief and governor were often combined, and when that happened the joint responsibilities were frequently bestowed on soldiers rather than on civilians. In the twenty years before the Crimean war turbulence on the frontier provided many instances of such fusion. Sir John Colborne (the future Lord Seaton), appointed lieutenant-governor of Upper Canada in 1830, was told to stay as effective governor-general when rebellion broke out in 1838. In South Africa, the sequence of Kaffir wars led to a succession of military governors, including Sir George Napier (1837–43), Sir Harry Smith (1847–52), and Sir George Cathcart (1852–4). In India, reeling from the aftermath of the 1st Afghan war and anticipating hostilities in the Punjab, Sir Henry Hardinge was appointed governor-general in 1844. All these men were military heroes of the Napoleonic wars, and they are not isolated examples.

The pattern continued into the twentieth century. Indeed, it was even reinvigorated by the impact first of global war and then of decolonization. As India faced invasion by Japan, Lord Wavell was created viceroy in 1941. After the war, with Malaya threatened by communist insurgents, Sir Gerald Templer was accorded joint military and civil powers in 1952, and so satisfactory was the outcome that similar principles were followed in Cyprus, with the appointment of Lord Harding in 1955. In 1978 the recently retired quartermaster-general, Sir William Jackson, went to be governor and commander-in-chief in Gibraltar; the example was repeated in 1993 when the outgoing chief of the general staff, Field Marshal Sir John Chapple, took up the same combined appointment.

Government by the army may have been deemed unconstitutional within Britain, but Britain saw nothing bizarre in imposing military government elsewhere. Britain has few examples of victorious soldiers turning their military reputations to advantage in domestic politics. In this respect the comparison with other countries is striking. But the tone of self-congratulation fails to take

account of the safety-valve provided by the empire. If the perspective is widened beyond Westminster, what becomes evident is not
how few generals were politically active after the two world wars
but how many. Allenby became high commissioner in Egypt in
1919, and Byng was governor-general of Canada from 1921 until
1926. Closer to home, French was lord-lieutenant of Ireland
between 1918 and 1921. After the Second World War, Alexander
went to Canada in 1946, Slim to Australia in 1953, and Bernard
Fergusson to New Zealand in 1962. The outstanding absentee in
this catalogue is Montgomery. He did not acquire a governorship
but he did involve himself in international politics—much to the
embarrassment of everybody but himself.

The heart of the empire was India, and it is with the subcontinent
that a closer study of these points must begin. India was won by the
sword and, as significantly, it was held by the sword.

The removal of France's challenge to British hegemony in India
by the end of the eighteenth century only fleetingly reduced the
primacy of military force. The combination of what was seen to be
the traditional basis of Indian princely authority with the pressure
to consolidate Britain's control over the subcontinent made efforts
to demilitarize redundant and even dangerous. The governor-general whom Hardinge succeeded in 1844, Lord Ellenborough, was
castigated for the bellicosity of his policies. He saw war not as an
instrument of policy but as identical with it; he recognized that the
army was the foundation of British prestige. Defending his record,
he reminded his critics that when he arrived in India in 1841 it was
to confront a crisis generated by the defeat of that army in Afghanistan. 'It seems to be forgotten that I found in India a defeated Army,
and an Empire in danger. It was in danger because acquired only
by Arms, by Arms alone could it be preserved, for in the heart of
the people it has no foundation.'[4]

Although Britain's military reputation was refurbished in the two
Sikh wars of 1845–6 and 1848–9, they did not obviate the outbreak
of the Indian Mutiny in 1857. The narrow margin by which Britain
then retained its hold on India confirmed the wisdom of Ellenborough's observations. Until 1912 every commander-in-chief in India,
with the exception of Kitchener, had served in the mutiny. Thus the
threats to British sovereignty that preoccupied them were not just

the external ones generated by the advance of Russia into central Asia; they were also internal.

The army therefore enjoyed a position relative to India's security—and hence to India's government—that found no parallel elsewhere in the empire. Francis Rawdon, Lord Hastings, who was governor-general from 1813 until 1822, and was himself a soldier, declared: 'The first duty of the government is to fix the amount of military force necessary for the maintenance of India; that adequate provisions must be made for its support, and that if reductions be then necessary to square the receipts and expenditure of India, it must fall upon the civil charges of government.' In Hastings's hands, military costs rose by almost 20 per cent in ten years, and when the Burma war of 1824–6 is included by 69 per cent.[5]

Such prodigality created tension between the governor-general in Calcutta and the Court of Directors of the East India Company in London. For the latter commercial success was being jeopardized by the pursuit of strategic security. Their efforts to bring military expenditure to heel were not aided by Hastings's decision to create a committee in India with the task of reviewing the instructions of the Court of Directors: the members of the committee included three civilians but eight soldiers.

The army was not only extravagant; it also appeared to be inefficient. In 1824 a mutiny at Barrackpore cast doubt on the quality of the officers of the Bengal army, suggesting that they put their own personal and political interests ahead of those of their regiments. What was particularly galling was that the officers of the Bengal army (as opposed to those of the other two presidencies, Madras and Bombay) received an additional allowance. Called half-batta, this was normally payable to soldiers on active service, but in the case of Bengal it was paid even while in cantonments. Harris's successor, Lord Amherst, undertook to deal with the half-batta issue, but he recoiled when confronted with the opposition of the commander-in-chief in India, Sir Edward Paget.

In 1828 the court appointed Lord William Bentinck to be governor-general. Bentinck's tasks included the pursuit of retrenchment and of efficiency. His earlier Indian career, as governor of Madras, had been broken on the back of another mutiny, at Vellore in 1806. He therefore had little faith in the Company's troops, and hoped to reduce the native infantry while expanding the numbers of British regiments. And almost his first action was to cut the half-batta payable to the Bengal army.

By the criteria of military sociologists—even if not according to the judgements of many of their contemporaries (including, of course, Bentinck himself)—the officer corps of the East India Company's army was more professional than that of the forces of the crown. P. E. Razzell's figures suggest that for the period 1805–34 76 per cent of the Company's officers could be described as middle class, as opposed to 47 per cent of the king's officers. There is, of course, a problem here with the terminology of class, and especially its appropriateness in the context of the early nineteenth century. C. A. Bayly has preferred, in describing the effects of prolonged warfare with France and in the empire, to speak of the re-creation of a *noblesse d'épée*.[6] The empire, like the army itself, provided a focus for the aspirations of the Scottish gentry. India was a path to promotion for ambitious officers; it was also a haven for the impecunious. 'An officer', Edward Ingram has asserted, 'went to India to make his fortune.'[7] Pay was therefore a much more important issue for the Indian army than it was for the British. Bentinck's assault on the half-batta question attacked the Bengal officers' sense of professionalism in two ways. Immediately it threatened their incomes: one manifestation of their professionalism was that these mattered. More widely, it challenged the political primacy which the army had created for itself. This sense of their position had been legitimized by Harris's governorship.

The opposition to the governor-general took three forms. First, senior officers snubbed him by refusing his invitations. Secondly, they elected their own council to present their arguments and to represent them with the Court of Directors in London. And, thirdly, the commander-in-chief, Lord Combermere, who as Stapleton Cotton had commanded Wellington's cavalry in the Peninsula, declined Bentinck's request that he support the governor-general and instead let it be known that he sympathized with the officers.

Bentinck had thus evoked a degree of professional solidarity so great that it spanned the divide between Company officers and the officers of the crown—or at least the most important of them, Combermere. In 1829 Bentinck feared mutiny. It did not come to

that, but Bentinck felt himself compelled to execute what amounted
to a coup in reverse. In 1833 he took over the post of commander-
in-chief himself, so combining in one person both civil and military
responsibilities.[8]

However, this gesture was not quite the assertion of civil authority over military that it might appear. Bentinck was, after all, himself
an army officer. What was at work, as in eighteenth-century Britain,
was fusion. The army in India was professionally aware and prepared to be politically active. But in India the effects of fusion did
not stop there. The consequence of having an empire dependent on
such an army was, as Douglas Peers in particular has argued,[9] to
militarize the government. Because the civil authorities recognized
their dependence on the army, and because they appreciated the
undesirability of separating the civil government from the military,
they colluded in what was happening. Despite the tensions in
Bentinck's relations with the army, ultimately they did not break,
and no other governor-general so tested that bond in the early
nineteenth century.

The most obvious symbol of this fusion was the Indian Political
Service. Essentially India's diplomatic corps, the political service
represented British government on the military marches and in the
border areas. M. E. Yapp has demonstrated that its members were
the key agents in promoting a forward policy on the frontiers of
India in the 1830s, 1840s, and 1850s. They had the monopoly of
local information; they tended to emphasize the dangers in their
own areas; and they were anxious to advance their own careers.[10]

In 1815 a bare majority of the posts in the Indian political service
were held by civilians. By the 1830s most were occupied by soldiers.
They had served briefly with their regiments, passed an interpreter's examination, and then transferred via the staff to the
political department. Thus the able and ambitious side-stepped the
drudgery of promotion by seniority, and while still comparatively
youthful gained appointments that were challenging and responsible.

But this created a problem of identity for the political service.
Were its members soldiers or administrators? Tensions and jeal-
ousies emerged between the politicals who were also officers and the
army proper. Although the former held ranks that were comparatively junior, they exercised powers that were comparable with

those of a general. Indeed, they could—by virtue of their status as politicals rather than as soldiers—use their political authority to override the military judgements of more senior soldiers, and so not infrequently direct the operations of a general.

The army's sense that the political department got it into scrapes because of the military immaturity of its members was exacerbated by the suspicion that some at any rate 'went native'. Long familiarity with the tribes that they were administering was reckoned to soften their martial instincts. Politicals, it was alleged, preferred to parley than to fight, so allowing situations to slide, and so rendering originally manageable situations extremely dangerous. In this interpretation the politicals jeopardized the military prestige on which Britain's supremacy in India resided.[11]

The 1st Afghan war produced a major crisis in this relationship. In the winter of 1841–2, General William Elphinstone withdrew his army from Kabul, through the gorge formed by the Kabul river, in the direction of Jellalabad. His command was massacred, almost to a man. The reputation for military invincibility, on which Britain's authority on the subcontinent was reckoned to rest, tottered. General William Nott, commanding the garrison at Jellalabad, was clear as to who was responsible. The fault was not Elphinstone's individually or the army's collectively. It was that of Sir William Macnaghten, the political officer in Kabul. Macnaghten had underestimated the wiles of the Afghans, and yet had been able to override the professional common sense of the soldiers. 'If a man is too stupid or too lazy to drill his company', Nott fulminated, 'he often turns sycophant, cringer to the heads of departments and is often made a Political, and of course puts the government to an enormous expense and disgraces the character of his country.'[12]

In December 1841, the very same month in which Elphinstone began his retreat from Kabul, Major-General Sir Charles Napier arrived in India. The subcontinent's mixture of politicized soldiers and militarized politicians was a heady one. The combination of the worries of internal security with unsettled and uncertain frontiers was potentially volatile. Napier's was not a temperament whose instincts would lead him to pour water on such combustible elements.

The Napiers were a family of military professionals. Charles's father George was a colonel; both his brothers—another George (the governor at the Cape) and William—were generals too. It was also a family of political radicals. George senior had witnessed revolution in America, France, and Ireland: he concluded that rebels rebelled for good reason. His convictions were reflected in the masterwork of one of his sons, William's *History of the War in the Peninsula*. The vivid prose of its six volumes helped generate in the British army a sense of professional self-esteem. But it also provoked furious controversy, embodying as it did William Napier's admiration of the French Revolution and of Napoleon, and his denigration of Britain's governments and of the aristocracy which dominated them.

In the 1830s William was asked to stand as a radical candidate for a number of parliamentary seats. He refused, but he spoke on behalf of the reformers, and especially of J. A. Roebuck. He advocated, as did Charles, the introduction of universal suffrage and of the ballot, and he favoured a shorter life for parliaments. In 1835, in a speech to a reform meeting in Bath, William declared that it was only a half-truth to aver that soldiers had nothing to do with politics. They had, he went on, 'nothing as a body, everything as individuals . . . if a soldier does not know and love the social happiness springing from equal and just laws, how, in God's name, is he to fight as the soldier of a free nation ought to fight?'[13]

William Napier's only employment after 1815 was as lieutenant-governor of Guernsey, a post he held between 1842 and 1847, and significantly another military-cum-political office. His primary importance in the present context is his self-appointed role as publicist and advocate for his brother Charles.

According to William, Charles's political education was developed in 1814, when he went to the Royal Military College at Farnham. As well as military science, Charles studied 'history, policy and civil government; adding for extension of matter, his own experience of British mal-administration in the colonies'.[14] Like William, Charles was critical of the aristocracy and of the Whigs. But for both of them, their radicalism was tempered by royalism. They wore their hair long (Charles cut a particularly disreputable figure on formal military occasions), thus marking out their identification with the Cavaliers rather than the Roundheads. Charles emphasized that, although he was an 'ultra-radical', he was not a republican. Indeed, he opined, 'my thoughts are strongly inclined to a good tough despotic government, *sufficiently bloody*, as the best of all forms of ruling. I would rather be cut down by O'Connell's guards in a rage, than be constitutionally strangled after a long speech from *The Mandarin*, Lord John Russell.'[15] The paradox in such apparently contradictory sentiments found some resolution in Napier's professional calling. Military imperatives justified absolutism: in war 'all history teaches the necessity of confiding command to a single man, that he may direct the warlike energy of the nation with full effect'.[16]

For eight years, between 1822 and 1830, Charles Napier was the British resident in Cephalonia, in the Ionian islands. This experience of colonial administration confirmed his faith in benevolent despotism. He set up schemes for improving the island—building roads, prisons, and hospitals—which he always prefixed, when describing them to others, with the possessive pronoun in the first person singular. Unsurprisingly his tenure of office was marked by controversy. In 1825 he was invited to head a Greek army in the war of liberation against the Turks: the suggestion came to naught when the Greek committee in London opted to channel its funds in other directions. More importantly, he fell out with his superior, Sir Frederick Adam, another soldier and lord high commissioner of the Ionian islands. His vendetta with Adam formed the leitmotiv of his first book on colonial administration, *The Colonies: Treating of their Value Generally—of the Ionian Islands in Particular*, published in 1833. A second work, *Colonization*, projecting an idealized form of colonial administration and prepared in unfulfilled anticipation of his assuming a governorship in Australia, appeared in 1835.

In 1839 Charles was appointed to command the Northern District in England. In many ways it was an extraordinarily courageous choice. His task was to confront the threat of Chartist insurrection. It asked Napier to set at odds his professional calling and his political convictions. His sympathies lay with the Chartists. Like them he detested the new Poor Law, and he could not forebear pointing out to the Whig government what he felt to be the proper political remedies to deal with the situation. Lord Fitzroy Somerset, military secretary to the commander-in-chief, Lord Hill, reminded

Napier that it was not Napier's job to speak out on political matters; his tasks were purely military. Napier took no notice.

None the less, Napier was undeniably successful in the tactics which he used to confront the Chartists' threat to use force. He did successfully distinguish between his professional commitment and his radicalism, and the combination of firmness and sympathy to which this gave rise provided the proof as to how inspired his appointment had been.

Napier's own rationalization of his position was published, extraordinarily enough, right in the middle of his tenure of the command. In 1840 he edited an English edition of Alfred de Vigny's *Servitude et grandeur militaires*. In his commentary, Napier gave an example in which a British general is called out to aid the civil power by dispersing a mob agitating for the repeal of the Corn Laws. As it happens, the general is also an MP who has proposed the abolition of the Corn Laws in parliament. The predicament of Napier's general is, to use the vocabulary of the previous chapter, a consequence of fusion. But the point which Napier was anxious to extract was that of dual control. Faced with this situation, the general behaves as a soldier because the army must obey the sovereign. If parliament commanded the army rather than just paid it, then the general who was simultaneously an MP would be under pressure to act out of his political persona rather than out of his military one. 'The essence of an army', Napier wrote, 'is *obedience* to its chief; the chief of the English army is the sovereign.'[17]

This seems unequivocal. But it was not. Neither in his career nor in his writing was Napier's commitment to obedience unconditional. In the preface to his edition of de Vigny, he made clear to soldiers that 'we obey, not from blindness, "not upon compulsion", but from reflection'. The scope for independent opinion implicit in this view was made explicit by Napier's observations almost 400 pages later. He argued that it was not the monarch but her ministers who were responsible to the people; as the army owes its loyalty to the queen, but the queen is left free of direct responsibility, the constitution—as interpreted by Napier—permits the army to arrogate responsibility to itself.[18]

Napier, like other politicized soldiers at other times and in other countries, squared these contradictions by reference to the 'nation'.

Because the pay of the army is voted annually by the House of Commons, Napier is at one stage able to identify parliament and the constitution with the people, and to conclude that 'the army is thus created by the breath of the nation, and dies at its will'.[19] But clearly the concept of dual control will not allow this identification to persist: it is at odds with Napier's earlier distinction between the parliament that pays and the sovereign that commands. The nation is amorphous, and its attraction to Napier lay precisely in this lack of precise form: it could become a projection to be shaped so as to fit the imperatives of the army.

Among the concluding remarks of Napier's edition of de Vigny was a diatribe against 'the despots of the East India Company'. He accused its directors of enriching themselves at the price 'of happiness among a hundred million's of people'; he declared—in sympathies redolent of his brother William's—a preference for the more beneficent despotism of Napoleon.[20] Thus the directors could not say that they had not been warned. The general who arrived in India in 1841 was fully politicized, with experience of colonial administration, and with radical views on domestic political reform. He was also an entirely professional general, committed to the army as a career, and anxious—as others had been before him—to seize his moment in the subcontinent to establish his military reputation.

Almost immediately after Napier's arrival in India, Lord Ellenborough was appointed governor-general. Ellenborough was convinced that the defeat in Afghanistan threatened to destabilize Britain's position in north-western India, and was consequently determined to restore the prestige of the army. He also shared the army's distrust of the political service. He therefore appointed Napier to the command in Sind, in what is today southern Pakistan, and placed the politicals in the area under him.

Napier's sense of natural justice was outraged by the rule of the amirs of Sind. However, the resident in Sind, Major James Outram, an officer of the Bombay army who had transferred to the political service in 1835, began to take the opposite view, arguing that the local population was happy under the amirs' rule. Outram was an Aberdonian of impoverished background, who had sought political employment on the north-west frontier in the conviction that there 'glory and honours . . . can be obtained', and that 'a man once placed there is sure to rise if he *deserves* to do so'.[21] Both Ellenbor-ough and Napier were not slow to note that the continuation of the amirs' government also served to buttress Outram's personal power. Both feared another set-back to Britain's military position gener-ated by the politicals' instincts for the appeasement of the tribes.

In February 1843 Napier took matters into his own hands and defeated the amirs at Miani. His lust for victory, his pursuit of the professional coping-stone of an independent command in war, had been assuaged: personal ambition drove Charles Napier just as surely as it fuelled James Outram. Sind was annexed and Napier was appointed its governor.

Outram was furious. He himself had initiated a forward policy in Sind. His umbrage at Napier's action therefore owed less to the policy which it represented, even if latterly he had distanced himself from that policy, and more to Ellenborough's subordination of the politicals. Immediately after the battle, Outram proceeded on an overdue leave. He paused in Bombay for a month, where he was fêted and where his sense of grievance was stoked. In May he arrived back in Britain, ahead of the official dispatches from either Ellenborough or Napier. Outram's criticisms were listened to with an avidity that reflected the preconceptions of their listeners. The Court of Directors was keen to call a halt to costly military activity on the frontier; the Whigs were looking for ways to discredit Ellenborough; the London press, fed by the editor of the *Bombay Times*, portrayed Napier as ruthless and vainglorious. On 29 August 1843 the Court of Directors passed a resolution to the effect that the annexation of Sind had been unjust and impolitic.[22]

Arguably, the issue might have blown over with the arrival of the official dispatches and with the cabinet's declaration of support for Ellenborough. But it was at this point that William Napier took up his pen in support of his brother, and in so doing fanned the flames of the dispute. Outram was stung into making further accusations against Charles Napier, including the historically powerful one that he had made war to make money. Charles himself, provoked by what he saw as a less than fulsome vote of thanks from parliament, called for a full parliamentary inquiry. Ellenborough, recalled from India in 1844, encouraged William to press on with the Napier version of events, *The Conquest of Scinde*, published in 1845.

Ellenborough's departure from India did not produce a dramatic change in policy. Hardinge, his successor, was not only Ellenborough's brother-in-law, but also a distinguished soldier. Fusion, or even militarization, persisted. Napier set about ruling Sind as he had done Cephalonia—as a benign despot. He installed a military government, excluding the politicals, and insisting that the people of Sind would only respect true soldiers who were themselves the conquerors of the land which they occupied. Civil government, he warned, would prove expensive, inflated, idle, and ill-informed.[23]

In 1847 Napier returned home. The bickering between the Napier brothers and Outram, and through him with the Court of Directors, rumbled on. More than age, controversy seemed to make it impossible that Charles could ever return to India. But in 1849 he did. Sir Hugh Gough's mishandling of the 2nd Sikh war obliged the Company to find a more able commander-in-chief in India before disaster overtook its army. Wellington was adamant that Napier was the only man for the job. The Court swallowed its collective pride and appointed him.

By the time Charles Napier landed in India for the second time Gough had fought the Sikh war to a bloody but victorious conclusion. Napier's task was not therefore the command in the field for which the Court of Directors had been persuaded to choose him, but the military administration of India. For Napier, reflecting the views of so many governors-general, this was not a subordinate issue in the consolidation of Britain's hold on the subcontinent, but its very pith and marrow. Without an efficient army that enjoyed primacy in the rule of India, Britain's position would be continuously vulnerable. What worried him above all was the possibility of mutiny in the native army.

His prescience in this respect was vitiated by two factors, or rather by two personalities. The first was of course his own, ambitious, sensitive to his own status, and quick to take offence. The second was that of the governor-general, Lord Dalhousie. A Whig, an aristocrat, and inimical to the military pretensions of the army in India, he possessed convictions that by their very nature would not have endeared him to Napier. At the outset both dutifully made the right noises about respecting each other and each other's sphere of influence. The truce did not last, and its main obstacle was Napier's

disillusionment as to the political importance of the post of commander-in-chief—'a very low-bred, miserable, sneaking, toad-eat-
ing post it is', he moaned. The nub, as he went on to complain, was
that 'I am not commander'.[24]

Napier wanted the army in India concentrated rather than scattered in small detachments. This would enhance discipline and
improve the opportunities for training. It would thus minimize
the likelihood of mutiny. However, it would also have the effect of
reducing the powers of the political officers, because there would be
fewer troops in each locality and therefore insufficient immediately
available for the furtherance of their independent objectives. The
corollary of concentration, as Napier's opponents in the political
service were not slow to point out, would be an increase in the
dominion of the commander-in-chief.

Napier soon discovered that he lacked the authority to alter the
disposition of the troops in India—that was in the hands of the
governor-general. 'The commander of 300,000 men', Napier
railed, 'can't move two companies out of danger without leave of
the civil power!'[25] Napier took to muttering behind Dalhousie's
back, and his disloyalty reached Dalhousie's ears. After a review
at Peshawar, Napier had said to one of his officers, 'Look at that
force: if I were not tied hand and foot by the G.G., by God! I would
be in the Khyber tomorrow, and in a fortnight I would be at Cabul.'
Dalhousie's response was, understandably, to tighten the reins on
his commander-in-chief, not loosen them.[26]

Within a year matters reached a head. Napier argued that the
recently conquered Punjab was insufficiently settled to be under
civil administration. It is not difficult to see in his approach evidence of his distrust for the politicals, of his preference for benevolent despotism, and of the example already set by his
administration in Sind. Dalhousie, needless to say, took the opposite
view.

Then, in April 1850, Dalhousie cut the allowances paid to the
sepoys serving outside their home presidencies in the Punjab.
Mutiny broke out. The issue touched many raw nerves in the
army of India—including that of batta but above all that of Britain's
ultimate dependence on force of arms to sustain its rule. Napier
could argue that if he had had his way and had been responsible for

the administration of the Punjab, the danger would never have arisen. But, now that it had, one course of action open to him was to disband the two most mutinous regiments. He feared, however, that that would not contain the disaffection, and so the position not only in the Punjab but also in northern India as a whole might be threatened. Instead, his solution was to restore the allowances which the governor-general had withdrawn.

Dalhousie was out of contact, away at sea for the benefit of his health. Napier insisted that he had to act quickly to contain the problem. Dalhousie did not agree, maintaining that Napier should have waited for his return before so openly revoking his decision. On 13 April 1850 the governor-general issued a public reprimand of his commander-in-chief.

Napier's response was to initiate a trial of strength with the governor-general, to pit the authority of the army against the primacy of the civilian administration. He sent in his resignation, gambling on it being refused. What he hoped would happen was that Wellington would continue to back him as he had in 1849, a support which had proved so efficacious with the Court of Directors. But on this occasion, the duke was unable to endorse his actions, and Napier's resignation was accepted. His bluff called, Napier took ship for home.

Given Napier's career and the views which he had expressed so forcefully and so frequently over such a long period, Dalhousie was surely right to see the issue of the allowances to the mutinous sepoys in a more general context. Publicly, Napier of course pointed to the immediate and specific danger with which he had to deal. But privately, and even not so privately, Napier had made clear his desire to expand the authority of the commander-in-chief in India, and his frustration at the subordination of military power to political. Dalhousie, Napier wrote to his sister on 15 May 1850, 'tries to act Coriolanus and only acts Tom Thumb'.[27] If Napier had been given the powers that he craved, both the governor-general and the Council of India would, in Dalhousie's view, have become superfluous.[28]

Charles Napier died in 1853. His final achievement was to pen and publish his last book, *Defects, Civil and Military, of the Indian Government*. The main thrust of this broadside was of course a

sustained criticism of Dalhousie's policy and a stout defence of
Napier's actions. But he also reiterated the principles espoused so
fervently by Hastings, Ellenborough, and others. 'In India', he
wrote, '*peace is never certain for a single day.*' Thus, he went on, 'An
Indian Commander-in-Chief may in a moment find himself, without preparation, responsible for the safety of the Indian empire.
His position is in no way like that of home Commander-in-Chief,
on whom events so sudden and so terrible cannot burst. Therefore
the former ought to have power commensurate with his vast
responsibility.'

Napier's demand seemed to be that soldiers should have political
authority: 'Many are the examples of danger from divided power in
war and the pernicious influence of civil authorities; and also of
military men invested with civil power—*politicals.*'

Napier referred of course to Macnaghten, Elphinstone, and the
1st Afghan war. But he blamed Sir Frederick Currie—the resident
in Lahore—for similarly irresponsible action in the Punjab in 1848.
And he concluded, 'with remembrance of my own narrow escape in
Scinde from the fatuous political, Outram. I resolved to maintain
firmly the integrity of military command while I could—when I
could not to resign. Woe to the country whose ruler employs
subordinates to advise, to suggest, to dictate about military
matters.'[29]

Napier's polemic therefore addressed two issues arising out of the
army's political position in India. The first was the power and status
of the commander-in-chief in relation to that of the governor-general. The second was the unity of military command in operations
when political officers were present. When he wrote his book it
seemed that Dalhousie had put a check on the political aspirations
of soldiers in India—that fusion had been replaced by subordination. But by the time that the fourth edition of *Defects, Civil and
Military* appeared in 1857, the mutiny that its author had prophesied
had come to pass. Dalhousie's policies were discredited, Napier
apparently vindicated.

The Indian Mutiny cast a long shadow over British rule in India.
It reaffirmed in concrete fashion, as no verbal reiteration by Ellenborough or others could have done, that Britain's hold on the
subcontinent rested on force of arms. Thus the issue of internal

order was to shape the constitution and distribution of the Indian army in the second half of the nineteenth century even more powerfully than it had in the first half. Indian officers and imperial strategists might increasingly emphasize the progress of Russia towards Afghanistan, but the more immediate and sudden danger remained within India itself. Napier's characterization of the commander-in-chief in India—'He lies down at night in peace, he wakes at daylight to fight a general action!'[30]—regained its validity and its vitality in 1857. His successors in the army of India would use both arguments to restate the case for the militarization of the government of India. They would prove more successful than Napier had been.

5

ARMY AND EMPIRE:
FROM KABUL TO THE CURRAGH

One of Charles Napier's most trusted subordinates in Sind was a future general, Abraham Roberts. Roberts had resigned his command in the 1st Afghan war rather than accept the advice of a political officer. The example was not lost on Roberts's son, Frederick. The latter followed his father into the army, as his son followed him. Frederick Roberts was therefore a scion of a professional family. He won the Victoria Cross in the Indian Mutiny, and his early career was stamped by that experience and by its message for Britain's military prestige in India.

Frederick Roberts became an icon of late Victorian Britain. Dubbed 'Bobs' by an adoring public, his diminutive stature was offset by a full white moustache. Like Gladstone's and Disraeli's, his image and reputation were the fruit of a flourishing press and the proliferation of illustrated papers. Like them too, he knew how to use his popular following for the achievement of political ends. The trajectory was set in the 2nd Afghan war, and in particular by the march of his column from Kabul to Kandahar in August 1880. His superior officer, Sir Donald Stewart, had probably accomplished more in his earlier march from Kandahar to Kabul, but it was Roberts's achievement that captured the public imagination, and legend proved more powerful than reality.[1]

Unlike Charles Napier, Roberts was not troubled by poor relations with his viceroy (as the governor-general had become after the mutiny). Lord Lytton, appointed to India in 1876, shared Roberts's enthusiasm for a 'forward' policy on the north-west frontier, and in particular for the creation of Afghanistan as a buffer between Russia and British India. Lytton's patronage ensured that Roberts was given command of the Kurram column, one of three preparing in 1878 for the invasion of Afghanistan. But, if Roberts felt confident in his relationship with the viceroy, he was less sure of the politicals. He requested that his final orders should make it clear that, 'while every deference is to be paid to opinions of politicals, the General in Command is alone responsible'.[2]

Almost immediately, Roberts found himself charged with usurping the political direction of the war. Ostensibly Britain's objective in invading Afghanistan was to get the amir to accept a British mission in Kabul. Lytton's strategy rested on an attempt to separate the tribes of the Khyber and Kurram valleys from Kabul. But it was not clear whether this was a long-term objective or a means of waging war. In January 1879 and again in May Roberts was accused of stampeding the policy of the British government: *The Times of India* reported that he had declared the Kurram valley to belong to the empress of India, not to the amir of Afghanistan.[3]

Then in September Britain's resident in Kabul, Louis Cavagnari, was murdered. Britain's task was now one of revenge. It declared its intention to buttress the authority of the amir against his more violent subjects. However, the amir was implicated in Cavagnari's death and the internal challenges to his authority threatened Afghanistan with disintegration. Britain needed a set of longer-term objectives even more urgently than before. But Lytton continued purposely not to give Roberts written instructions on political matters, 'in order that your action on *reaching* Kabul may be *perfectly unfettered*'. He urged Roberts 'to strike terror, and to strike it swiftly and deeply; but to avoid a *"Reign of Terror"*'.[4] Precisely how many men Roberts executed on his arrival in the capital, and how well proven the accusations against them, cannot now be firmly established. That the proceedings were arbitrary in their manner and excessive in their number is, however, not in dispute.[5] Lytton had reminded Roberts that the long-term objective was conciliation. Roberts had made that impossible, and it is hard not to see in his actions both a working-through of the legacy of the Indian Mutiny and a desire to deepen the Afghan war so as to further his own career and the objectives of the 'forward' policy on the frontier.

In the storm that followed it was clear that Roberts was still learning how to manipulate the press. At the outset of the campaign he had expelled the war correspondent of the London *Standard* for writing dispatches critical of his operations. Now the Indian papers complained that only one trusty hack, Howard Hensman, was permitted entry to Kabul, so as 'to write a pleasant and safe account of the affair'. The viceroy could not let Roberts's political freedom continue untrammelled, not least because the excesses which followed were in danger of undermining support for the policy which they were designed to promote. In January 1880 a civil servant, Lepel Griffin, was appointed Roberts's political chief of staff.

Roberts's sarcastic letter of acknowledgement made no secret of his distaste for Griffin as an individual and for the new arrangements in general. But his frustration really became evident in May 1880, when Sir Donald Stewart arrived in Kabul. As Roberts's senior, Stewart became commander-in-chief, so stealing Roberts's thunder. Furthermore, Stewart was entrusted with joint political and military control. Roberts's solution was to change tack. He now argued that political control should be divided from military, and the former vested in himself and Griffin because they were much more familiar with the local situation than was Stewart. His preferred option was of course to have political affairs in his own hands, but when he did not get that he was happy for them to be given to Griffin rather than have all authority reside with Stewart. When he proved unable to get even this device accepted, he began to threaten resignation—despite the fact that he was a serving officer engaged in the middle of a campaign on hostile territory.[6]

Then, on 28 July 1880, came the news of the defeat of a British brigade at Maiwand, in southern Afghanistan. This area had been under Stewart's overall command, but when he had marched north to Kabul he had separated its political from its military control. Roberts and others were quick to see the parallel with the 1st Afghan war, and to attribute the blame for Maiwand not to any failings on the army's part, but to the politicals and to divided political and military authority.

Roberts himself set off from Kabul on his celebrated march to Kandahar, whither the worsted brigade had retired. This time he managed the press with extraordinary adroitness, the silence which fell over his column for three weeks heightening the impact of the news of its arrival. Roberts then used this reputation to push for the unification of political and military control in southern Afghanistan, and to threaten his resignation when he did not get his way. His fame now made it impossible for his demands to be resisted, and he duly received the chief political authority in southern Afghanistan.[7]

The 2nd Afghan war developed and honed Roberts's political skills. He exploited his influence with Lytton to maximize his own powers not only by the derogation of his professional rivals but also by playing off the viceroy against the politicals. He learnt to be respectful of the press, and how to use it to increase his popularity, and so gain leverage over the government. Finally, the war made clear his commitment to the Conservative party. The executions in Kabul were used by the Liberals to discredit Lytton's 'forward' policy in Afghanistan in the run-up to the 1880 election. Roberts viewed the prospect of a Liberal victory as a 'calamity'. Aware that he was—not least because his services had been political as well as military—'specially obnoxious to powerful sections of the Liberal party' (to use Lytton's words), Roberts was careful to cultivate the Conservative leaders, Lord Salisbury and Lord Randolph Churchill.[8]

Roberts returned to London at the end of 1880. In February 1881 he was accorded a celebratory dinner at the Mansion House. Although advised against it, he used the opportunity to deliver a major speech attacking the army reforms carried through by the Liberals when Cardwell was secretary of state for war. Roberts condemned the two pillars of the Cardwell scheme—short-service enlistment and linked battalions. Soldiers in India, he said, needed the stamina and maturity acquired by long service. The effect of the linked battalion system had been to send out drafts of men who were too young and too untrained.

Roberts was now dubbed Britain's 'only other general'. One of his auditors at the Mansion House was the rival implicit in this description, Garnet Wolseley. In 1871, as assistant adjutant-general, Wolseley had hitched his star to Cardwell and Cardwell's scheme. Stung by Roberts's criticisms, he promptly penned a reply for the March 1881 issue of the *Nineteenth Century*. The language was intemperate. He castigated the critics of short service as chronic grumblers and whining pessimists: he declared that 'all armies and navies are naturally conservative in their tendencies, and consequently view with suspicion any changes effected in their organization by a Liberal Government'.[9] The rivalry now assumed a form that was concrete, that went beyond the competing personal ambitions played up by the press, and embraced divergent policies.

Like Roberts, Wolseley was the son of a soldier and a true professional; brought up by a penniless widow, he was given a free commission in the 80th Foot. Like Roberts, he had learnt to manipulate the press. Roberts established his style and reputation in Afghanistan; for Wolseley the formative campaign was in Ashanti in 1873. The expedition was reported for the London *Daily News* by J. F. Maurice, his private secretary, and his military secretary, Henry Brackenbury, published the definitive two-volume account in 1874. Wolseley told Lord Melgund that 'the press has become a power which a man should try to manage for himself; that it is an influence which one cannot deny, and therefore should try to make one's own'.[10] But Afghanistan and Ashanti also pointed to a cardinal distinction: Roberts acquired his reputation in India while Wolseley made his in Africa. Each continent shaped a different strategic outlook.

Roberts's appreciation of imperial defence was framed by the security of the north-west frontier. He wanted an army whose bulk was concentrated in India on a permanent basis. His strategy was a precursor of the 'continentalism' of 1914, only the continent was Asia, not Europe. Wolseley, on the other hand, reckoned that the army should be based at home. His own campaigns—from Ashanti in 1873, through to South Africa in 1879, Egypt in 1882, and the Sudan in 1884–5—had been mounted from Britain. His strategy was therefore global and amphibious: it rested on the vulnerability of British seapower for its emphasis on home defence and on the navy's strength for its capacity for rapid deployment. It aimed to counter Russia not on the borders of Afghanistan but in the Mediterranean and the Middle East.[11]

These issues—long service versus short service, India versus Britain—were professional matters. Furthermore, each of the two generals had adopted their party political loyalties for professional reasons. Roberts favoured the Conservatives because they were committed to a 'forward' policy in India; Wolseley supported the Liberals because their reforms promised a modernization of the army. The difference in political loyalties, made so public in 1881, seemed to complete a polarity that ensured that each general was the opposite of the other.

In 1884 Wolseley set off on what proved to be his last campaign— the bid to rescue Gordon, besieged in Khartoum. The mission symbolized this combination of professional virtues with political loyalties. Wolseley's methods were those acquired in his earlier wars: above all, he opted to follow the Nile, the longest route to Khartoum, in emulation of his very first triumph, the Red River expedition of 1870. If successful, he would not only consolidate his own military reputation, he would also save Gladstone's Liberal government.

He failed. His painstaking techniques appropriate to other circumstances and other climes were not necessarily the right ones for the Sudan: most importantly they lost time. The responsibility was of course not only Wolseley's. Gordon brought about his own downfall: here was another imperial soldier resisting ministerial direction in an ambitious and vain (in both senses of the word) effort to shape policy. But by his death Gordon ducked any blame: he became a hero rather than a scapegoat. Wolseley could not bring himself to accept that the relief expedition was not the triumphant culmination of all that had gone before. He concluded that Gordon's death was not his fault but Gladstone's. Professional *amour propre* was more important than the loyalties of party politics.[12]

Wolseley's method of self-exculpation was to manipulate the tensions of dual control, to exploit the army's ability to play off the crown against the prime minister. Queen Victoria did not like Wolseley much: neither she nor the duke of Cambridge had looked with favour on his association with the Liberals. But she liked Gladstone even less. 'It is very disheartening to the soldiers of this army', Wolseley wrote to the queen,

to find that Mr Gladstone and all his colleagues have completely ignored all the toil they endured without a murmur on the river, all the fighting marches under a burning sun in the desert, and all the severe fighting they have had. It is very ungracious on Mr Gladstone's part, seeing that it is his fault that all these trials have been endured in vain, and all our dead comrades killed to no purpose. He might have said a few cheering words to these soldiers: they would have cost him nothing, and even from him, unpopular as he is in the army, they would have been valued. However, if the Queen is satisfied with the conduct of her troops, I don't think our men care very much what Mr Gladstone may think of them: *they* certainly don't think much of *him*.[13]

Two factors cemented Wolseley's divorce from the Liberals.[14] First, Wolseley, like many officers, came to see Gladstonians as 'little Englanders', inimical to the empire with which the army was increasingly identified. Secondly, as 'new liberalism' made headway in reorientating the party towards a greater commitment to state intervention, Liberals stood accused of socialism, of sapping the moral fibre of the nation and so contributing to the decadence which made the task of imperialism that much harder to sustain. At the junction of both imperial and domestic elements was the Liberal commitment to home rule for Ireland. In 1886, the year when the first home rule bill was introduced, Wolseley gave it as his view that Irish home rule would lead to the breakup of the empire. In 1893, the year of the second home rule bill, when Wolseley was commander-in-chief in Ireland, he continued to play off the crown against the government, fearing that 'the worry of the Home Rule Bill would kill the Queen. If so, Mr Gladstone will have another, and a very great sin to answer for.'[15]

Two years later, in 1895, the army's pursuit of its professional objectives, and the politicization which flowed from them, reached a climax. On 21 June, the Liberal government—from whose leadership Gladstone had now finally retired in favour of Lord Rosebery—was defeated in what a later Liberal leader was to describe as 'a snap vote in a thin House on a side issue'.[16] This is the customary interpretation of the so-called 'cordite vote'. But it underestimates its relevance to the present context: however unsurprising the Liberal split, it is still significant that the division occurred over the army estimates. The issue at stake was whether the army had sufficient munitions to mobilize three army corps as the secretary of state for war, Henry Campbell-Bannerman, maintained. Imperial defence was capable of determining party loyalties: professional lobbying had its effect. Moreover, the consequence of the Liberal defeat was that Wolseley was appointed commander-in-chief in succession to the duke of Cambridge. The Liberals had had it in mind to give the post to Redvers Buller. The Conservatives, anxious both to break the royal connection with the army and to implement the recommendations of the Hartington commission, were keen to scotch rumours that the choice might fall on the duke of Connaught.

At one level, the selection of Wolseley was entirely predictable: he was the most distinguished and most senior of late Victorian generals, and his antipathy for the old order and the duke of Cambridge would mark a new departure. On another, it was surprising. Wolseley had been rewarded with a viscountcy in 1885. His peerage had given him a parliamentary platform for his professional views. In 1888 the House of Lords had become the focus for a confrontation between Wolseley, then adjutant-general, and Salisbury, the same prime minister who in 1895 would appoint him commander-in-chief.

Wolseley had made remarks at a dinner which had then been reported, as he no doubt intended that they should be, in the *Daily Telegraph*. According to Salisbury, Wolseley had said:

The answer to the question why the Army and Navy are not as strong as they ought to be is to be found in the system of our government by Party—the curse of modern England which is sapping and undermining the foundations of our country—which is depriving our statesmen of the manly honesty which was once their characteristic. What do we see when any new Administration comes into Office? What directly takes place? It is the same with all Parties. The first thing is an endeavour made by the Minister in Office to obtain some claptrap reputation by cutting down the expenses of the Army and Navy, and if he is able to produce an Army or a Navy Estimate which represents in some degree a smaller sum than that of his Predecessor, he plumes himself upon the victory he has gained.[17]

Salisbury's objection was to the practice of 'distinguished authorities on military affairs' making 'statements against the Government under whom they serve, and making them in places where they cannot be answered'.[18] The *Daily Telegraph* had referred to Wolseley not by name but as the 'highest military authority'. Despite this implied insult to his own status as commander-in-chief, the duke of Cambridge none the less defended Wolseley's right to air his views on professional matters in public places, and specifically in the House of Lords. Salisbury, on the other hand, clearly subscribed to the view that the government's professional advisers should give their opinions in confidence.

Wolseley was not chastened. Writing to his wife two years later, he looked forward to the day when, 'the licence of democracy & socialism will be conquered by the sword, and succeeded by cruel military despotism. Then it will be that the men of talk will give way to the men of action, and the Gladstones, Harcourts, Morleys & all that most contemptible of God's creatures will black the boots of some successful Cavalry Colonel. A new Cromwell will clear the country of these frothing talkers, & the soldiers will rule.'[19]

These were not sentiments confined to the confidences shared between husband and wife. In more coded but published form, they provided the salient themes of a two-volume life of Marlborough written by Wolseley in 1894.

Wolseley's biography was essentially a political life rather than a military one: it ended with the accession of Queen Anne, before his hero had dazzled Europe with his triumphs in the war of the Spanish succession. The central conundrum, therefore, was Churchill's duplicity in 1688: his betrayal of James II through his conspiracy with William of Orange. Wolseley did not duck the issue. Churchill was 'driven to treason against his King by the King's treason against his country, and driven to leave the master he had long and faithfully served by that master's flagrant betrayal of his sacred trust'. He was therefore 'working for the real interests' of the country, doing what he saw to be his 'duty to his God'. Churchill was 'a patriot, not . . . a mere conspirator'.[20]

In what followed it becomes clear that Wolseley's political thought was akin to that of another reforming general whose reputation was made in the colonies—Charles Napier. Like Napier, Wolseley stressed obedience to the crown, but also emphasized that the obedience could not be unconditional. Marlborough, after all, was 'more than a mere soldier, owing military obedience to his Sovereign before all things. He was a power in the country.'[21] But this freedom of choice was true not just of Marlborough; it also applied to the rank and file. Here Wolseley, like Napier, reflected the regimental paternalism and the professional solidarity which prolonged overseas service nurtured, and which was at odds with the contemptuous and contemptible attitudes of Wellington towards his men. The soldier, Wolseley insisted, must identify with the cause for which he is called on to fight.

Although the British soldier is a volunteer, he is no mercenary, no mere hireling who will fight in any cause, be it just or unjust, for the Prince or Government who pays him. He is not a mere piece of machinery to be wound up like a clock, or regulated like a steam-engine or a spinning-jenny. He has not only a body to be shot, but he is endowed with the same feelings and the same love of life as other people, and with the same respect and enthusiasm for a righteous cause as the best in the land. His heart—for he, too, has a heart—must be in the contest, and if it be not, there is little to be got from him. The Government or the General who counts upon the British soldier to fight well in an unrighteous and unjust cause, relies for support upon a reed that will pierce the hand which leans upon it.[22]

The crucial question implicit in all this is the identity of the agency which is to judge the causes which are righteous and just. Wolseley made clear in his life of Marlborough, just as he had made clear in the 'leaked' speech of 1888, that that agency was not necessarily the political party currently in power. 'Marlborough', Wolseley told his readers, 'never sought to perpetuate power in the hands of any one set, but employed men of both political parties, only considering the advantage of the nation.'[23] Highly developed in Wolseley, therefore, were those feelings against party politics which soldiers were apt to cite as evidence of their political neutrality, but which in essence constituted a contempt for parliamentary government.

The soldier's loyalty was to the nation. In 1688 Churchill's duty to his country had outweighed his duty as a soldier. Churchill's rejection of party was driven by the fact that his 'policy was essentially national'.[24] But the hole in Wolseley's argument, as it was the hole in Napier's, was that he provided no definition of what constituted the national interest: this remained unspoken, locked in the mind of the soldier, and therefore readily adapted to the political needs of the army.

Wolseley lauded the 1688 revolution for its establishment of Protestantism and its guarantee of ancient rights and liberties; he even drew attention to the Mutiny Act and its stipulation that the army's existence was conditional on annual approval by parliament. But he was careful to resist Macaulay's interpretation of the significance of these events. The army, through Churchill, had been the key arbiter—not the Whigs or Tories. And Churchill 'had no strong theories about liberty'. 'The bent of his mind was towards a mild and beneficent despotism.'[25] If these were Marlborough's views, they were also Napier's. Wolseley's letter to his wife suggests that they were his too.

Wolseley kept parliamentary silence while he was commander-in-chief. Not until 1901, when he was in retirement, did the friction between him and Lansdowne, as secretary of state for war, flare into public life—and even then Wolseley might reasonably claim that the personal dimension in the exchange was not his initiative. But it would be wrong to conclude that, just because Wolseley felt precluded while in office from public utterance in the Lords, he therefore felt that his office required him to eschew any expression of his views on the politics of the army.

In 1898 he wrote the foreword to one of the most influential texts of late Victorian or early twentieth-century British military history, Colonel G. F. R. Henderson's *Stonewall Jackson and the American Civil War*. Henderson's own confusion on the subject of civil–military relations and the political control of operations has been referred to in Chapter 1. Henderson's reflections constituted the basis for the second of the two lessons (the first concerned the need for an efficient and sufficient army) which Wolseley derived from the book. Wolseley concluded that 'to hand over to civilians the administration and organization of the army, whether in peace or in war, or to allow them to interfere in the selection of officers for command or promotion is most injurious to efficiency; while, during war, to allow them, no matter how high their political capacity, to dictate to commanders in the field any line of conduct, after the army has once received its commission, is simply to ensure disaster.'[26]

In the following year the British army encountered disaster—at Stormberg, Magersfontein, Colenso, and Spion Kop. The Boer war called into question Wolseley's tenure of the post of commander-in-chief, and it discredited two of his most distinguished acolytes. Sir William Butler had gone out to South Africa as commander-in-chief in 1898; the absence of Sir Alfred Milner on leave had also bequeathed him the post of acting high commissioner. Charged with political and military responsibilities at a time of high tension in Anglo-Boer relations, Butler had sought to appease rather than to confront. To him was passed some of the blame for Britain's lack of military preparedness. The victim (if that is not too strong a word) of that unreadiness was Butler's successor as commander-in-chief in South ' Africa, Redvers Buller, another of Wolseley's protégés.

The opening battles of the Boer war therefore effectively smashed 'the Wolseley ring'. In 1873 Wolseley had formed a specially selected staff of 36 officers for the Ashanti campaign. Dubbed at first 'the Ashanti ring', this small group acquired its later title as its dependence on Wolseley and his patronage became evident from his subsequent campaigns. Its officers formed the core on which Wolseley drew when he went to South Africa in 1879 and Egypt in 1882. Inevitably those soldiers whose careers were forged not in Africa but in India were christened 'the Roberts ring'.

The existence of the 'rings' is evidence of the political awareness of the officers of the late Victorian army. The 'rings' developed the capacity of officers for faction and intrigue, and the divisions they created within the officer corps gave the politicians some opportunity to exercise leverage within the army. But they were not as tightly knit as the rhetoric or the press suggested. As the careers of their members blossomed, rivalries within the rings could become as acute as those between the rings. Furthermore, postings occurred that cut across the rings. In 1884, Ian Hamilton, who enjoyed Roberts's patronage after soldiering in India, served under Wolseley in the Sudan. In 1899, when Roberts superseded Buller in South Africa, he still retained Wolseley's man as his commander in Natal. Most striking of all, the officer who to all intents and purposes succeeded to the title of 'our only general'—Horatio Herbert Kitchener—did not fit securely into either of the two rings.

Kitchener was, like Roberts and Wolseley, the son of an army officer who rose to the top through colonial warfare. He served under Wolseley in Cyprus, Egypt, and the Sudan, but was often at odds with him. In 1896–8 he effectively triumphed where Wolseley had failed, reconquering the Sudan and gaining his first experience of the fusion of political and military responsibility as its governor-general. In 1899 he became Roberts's chief of staff in South Africa, and in 1900 succeeded him as commander-in-chief. His strategy for guerrilla war, with its use of scorched earth tactics and its establishment of concentration camps, combined military skills with political imperatives more obviously than did many colonial operations. In 1902 he was appointed commander-in-chief in India. It was his first brush with the subcontinent. Although South Africa had, ostensibly, made him Roberts's man, his views on the defence of India and on its military administration were at variance with those of his erstwhile superior. Above all, he rejected the legacy of the mutiny: the Indian army could be safely expanded, and its focus should be on external defence against Russia, not on the danger of internal disorder.

Before he left for the subcontinent, Kitchener told a friend of the current viceroy, Lord Curzon, that he would use 'the whole of his popularity and prestige to dominate the next viceroy'.[27] His desire to expand the authority of the commander-in-chief was thus made clear in advance. Kitchener was about to take up the cudgels which Napier had laid down after his battle with Dalhousie.

In London, Wolseley now believed that the commander-in-chief in Britain should be ex officio minister of war—a permanent unelected member of the cabinet. Thwarted in this aim, he announced in the Lords, during the course of his exchange with Lansdowne in 1901, that 'it should not be competent for the War Minister, or indeed the Cabinet, to give a final decision adverse to a proposal deemed of such grave importance by the military expert'. The commander-in-chief should have the right to make his views public whenever the occasion arose, and the public should be invited to judge the merits of the respective cases put by the political and military chiefs.[28] Wolseley's demands, while of a piece with his political thought, were in stark contrast with the ebbing powers of the commander-in-chief. The difference between his failure to expand the authority of the commander-in-chief at home and what was about to occur in India could not have been more striking or more ironic. While politicians found the sort of Bonapartism advocated by Wolseley in London unacceptable and even absurd, they were prepared to condone a very different outcome in the colonies.

Kitchener's principal bone of contention was the status of the military member of the Council of India. The military member was an officer junior in rank to the commander-in-chief, but was an ordinary member of the Council of India whereas the commander-in-chief was an extraordinary member. The military member was the official channel of communication between the viceroy and the commander-in-chief; he also controlled the supply and finance of the army. Only operations fell outside his purview. Kitchener objected to having his orders subject to criticism or correction by the intervention of junior officers. Curzon, while he had no great affection for the system, was pushed into defending it by the vehemence of Kitchener's attacks. Divided military authority at least had the virtue of ensuring civilian domination. And, as previous incumbents of the office had shown, the combination of personal prestige and common sense could make the arrangements workable.

To get his way, Kitchener used many of the same political devices as those employed by Wolseley and Roberts. In H. A. Gwynne, editor of the *Standard*, and Charles à Court Repington of *The Times* he had two firm supporters. Repington had served on Kitchener's staff in Egypt, but his military career had been cut short by marital complications. Thanks to Gwynne and Repington, Kitchener could ensure that his opinions appeared in the newspapers on the same day as they were tabled for confidential discussion by the Committee of Imperial Defence. In this way he could use his reputation as an invincible colonial commander to mobilize a public following. It was a pressure to which the Conservative government of Arthur Balfour was extraordinarily vulnerable between 1902 and 1905. Itself divided over tariff reform, it could not afford a public parting of the ways with a national hero.

But these direct methods were not the only ones used by Kitchener. An appetite for intrigue and subterfuge was more highly developed in him than it was in Roberts or Wolseley. He had, in addition to his journalistic contacts, three indirect routes by which he could bypass the formal hierarchy of communication. The private secretary to the secretary of state for war was Captain Raymond Marker. Kitchener encouraged Marker to believe that he—Marker—had been jilted by Lord Curzon's sister on the initiative of the Curzons; through his correspondence with Marker, Kitchener relayed his personal opinions—without any countervailing mediation—direct to the secretary of war. He had a similar network in the India Office: General Sir Edward Stedman, military secretary at the India Office, passed on Kitchener's personal views to the secretary of state for India. Thirdly, Kitchener maintained a regular exchange of letters with Lady Salisbury, whose husband was Balfour's cousin. Through her, he had a spokeswoman with the prime minister. Thus Kitchener sustained contacts with the three members of the cabinet most closely involved in the defence of India, and did so on terms which cut out both the viceroy and the government of India, leaving them oblivious of what was being said behind their backs.

The Conservative leaders did not listen just because of their electoral vulnerability. Balfour in particular was deeply worried by the threat to India from Russia. It seemed that the only feasible scheme for the defence of the frontier was a pre-emptive thrust into Afghanistan, with a view to meeting the Russians there rather than in India. What was needed, therefore, was an embryo field force, concentrated under the commander-in-chief. Kitchener's case for the centralization of the Indian command, like Napier's over fifty years previously, had a strategic rationale.

Curzon's slowness to appreciate the depth of the crisis with which he was confronted was the fruit of deliberate deception. Kitchener wrote in one sense to London, and in another to Curzon. However, it was also a consequence of this intermingling of operational necessity with peacetime usages. Curzon emphasized the practical objections to Kitchener's proposals rather than the constitutional. The concentration of too much power in the hands of one man would make for inefficiency: a commander-in-chief of capabilities less considerable than those of Kitchener—and arguably even Kitchener himself—would not be able to cope.

The first assertion of constitutional norms did not appear until June 1904 and came not from Curzon but from Lord Ampthill, deputizing for the viceroy while the latter was in London. 'Lord Kitchener's proposal', Ampthill pointed out, 'is in effect to revert to the system of combining executive and administrative functions under one head, a system which no longer exists in any other army in the world, which India abolished fifty years ago and which has at last been finally and emphatically condemned in England.'[29]

Curzon's return to India was delayed until the end of 1904 by the illness of his wife. Any doubts about the need to see the issue in constitutional terms were now swept away. In January 1905 Kitchener consulted several senior generals in India over the reform of the subcontinent's military administration. He was careful to canvass only those officers likely to agree with his point of view. Curzon complained that such behaviour was unconstitutional, as the entire issue was one for the government of India. Kitchener replied that he had gathered these opinions solely for his own use, and that they would remain private to him. In fact he sent them to Lady Salisbury, and from there they found their way to the India Office, the Committee of Imperial Defence, *The Times*, and the prime minister. In London, Curzon's point of view went almost unheard.

Kitchener's anxiety that Curzon should not appear a 'martyr to the maintenance of constitutional government'[30] was a reflection of his own greatest vulnerability. By March 1905 Curzon was speaking of the creation of a 'military autocracy' and of a 'military despotism'.[31] Now that the constitutional issue had been raised, Kitchener replied in kind. With extraordinary effrontery, he accused Curzon of wanting to combine civil and military government in his own hands, and compared him with Napoleon.[32]

Curzon had shifted tack too late. At the end of May a compromise was hammered out which constituted a moral victory for Kitchener. The post of military member was (in Curzon's word) 'disembowelled',[33] and renamed that of military supply member: its functions were primarily civil and its incumbent could not veto the proposals of the commander-in-chief. The fact that there was a military supply member at all was a contrivance to stop Curzon from resigning. Curzon regarded the scheme as unworkable. Extraordinarily he managed to get Kitchener to agree, and by June it seemed that the post of military member was to be revived. Curzon suggested Major-General George Barrow, a rising star, as an incumbent of distinction sufficient to restore some lustre to an office that had had much of its gloss knocked off by recent events. Kitchener seemed to accept this nomination. But the commander-in-chief then informed the secretary of state for India, St John Brodrick, that Barrow was unsuitable. Brodrick told Curzon to consult Kitchener as to his preference. Curzon bridled at the notion that the military member would be the commander-in-chief's man, and suspected that Kitchener wanted a 'dummy' in the job so that even if the post of military member survived it would remain emasculated. It became clear to him that the cabinet was prepared to see him go rather than confront the departure of the commander-in-chief.

Curzon resigned in November 1905. A soldier had toppled a civilian governor. One of Kitchener's biographers, George Cassar, believes that the quarrel was 'due less to a difference of opinion over dual control than to a clash of seismic wills'.[34] At the outset Curzon may indeed have been more concerned to find the most efficient solution rather than to protect civilian prerogatives. However, Kitchener's own behaviour disabused him of such naïvety. Curzon learnt that it was not the army's lack of political instincts that precluded military interventionism but the maturity of the political culture within which the army had to operate. Thus Curzon's most recent biographer, David Gilmour, is surely right to come to precisely the opposite conclusion from that formed by Cassar: Curzon resigned on a matter of principle.[35]

By 1909 all military administration in India was centralized on the commander-in-chief, and it was on him that the Council of India was entirely dependent. Curzon was proved right. The load proved too great for one man, and the poor performance of all the expeditionary forces mounted by India in 1914–15 could with reason be laid at the feet of Kitchener. Nor was it clear what the functions of the commander-in-chief should be now that hostilities had broken out—albeit not in Afghanistan as Kitchener or his supporters had imagined a decade previously. Should he remain in India as a sort of chief of staff, orchestrating the army's expansion, equipment, and deployment, or should he command it in the field? Kitchener had anticipated he would do both: in reality, as he was himself learning in London (or, as his critics would have it, failing to learn), the functions of administration and of command had to be shared in the age of the mass army.

At the very same time that the post of commander-in-chief in India was being elevated and its responsibilities expanded, that of commander-in-chief in Britain was finally being abolished. Extraordinarily, the parallel—or rather the contrast—was rarely drawn. The difference between what was acceptable in India and what was acceptable in Britain was so obvious as not to need explicit formulation.

Most surprising of all was the position of Lord Roberts himself. Roberts had of course been commander-in-chief in India between 1885 and 1893. In 1904 as a direct consequence of Lord Esher's report, he was unceremoniously ousted from his job as commander-in-chief in Britain, and the post itself abolished. But Roberts did not respond to these indignities by being drawn into Kitchener's intrigues. He saw no need for the abolition of the post of military member of the Council of India. In 1905 he reminded Curzon that many great commanders had not controlled in peacetime the services essential to the armies which they had then led in war. Alexander, Julius Caesar, Frederick the Great, and Napoleon had, 'but they were actually, or virtually Dictators as well as Commanders'.[36]

However, Roberts's pragmatism on the politics of high command should not be construed as evidence of a more general neutrality on matters of political significance. The last decade of his life was devoted almost exclusively to political activism. His involvement in two issues in particular—first conscription and then Ulster—showed how effectively the combination of imperial soldiering and of its rewards politicized Britain's senior officers. The House of Lords became the platform from which Roberts could project himself into popular politics, just as his presence in the upper house could confirm its members in their view that they were true representatives of national feeling.

Roberts remained a member of the Committee of Imperial Defence until November 1905. But he then resigned in order to take up the presidency of the National Service League. The case for conscription rested on the needs of home defence. Formally speaking, the National Service League was non-partisan, but by 1909 Roberts's objective was to make conscription Conservative party policy. The principal block to this objective was the fact that Balfour and the Conservative front bench saw the navy as Britain's best protection against invasion; for them the army's task was the defence of India. As Roberts himself had pointed out in 1881, India needed an army of long-service soldiers, not of conscripts.

The case against conscription was further undermined by Haldane's reform of the auxiliary forces. National service and the

Territorial Force were different solutions to the same problem. Roberts set out to show that the Territorials would be adequate only to meet raids, but would not suffice to counter the large-scale invasion which he now believed was possible. Although he failed to persuade Balfour, he did convince some backbenchers and also many of his Conservative colleagues in the Lords. Under Roberts's aegis, the League mobilized the press, publishing scare stories of fictional invasions; despite his age Roberts himself showed an extraordinary receptivity to new technology, making a gramophone recording extolling the League's case.[37] Between 1909 and 1912 the membership of the National Service League tripled, to almost 100,000. In March 1910 the League claimed it had the sympathy of 105 MPs, of whom only three were Liberal. The effect was to achieve Roberts's objective—to identify compulsion with the Conservative party. But this was still an association that Balfour and his colleagues resisted. Electorally it promised disaster; strategically it would be pre-empted by a stronger navy.[38]

Conscription was not only an issue of great significance to party politics. It was also a matter of supreme professional importance to the army. It was not a subject on which the army was united. H. O. Arnold-Forster gave it as his opinion that the Army Council was in favour of conscription from its inception in 1904.[39] But, even if this was true then, it became less true under Haldane. The army's younger and more progressive soldiers, including Sir John French, Sir Douglas Haig, and Sir Ian Hamilton,[40] were impressed by Haldane and became his devoted adherents. They resisted the imputation that senior serving officers were adherents of the League muzzled by professional obligations. In 1910 Haldane published what he represented as an in-house memorandum against conscription, called *Compulsory Service* and written by the adjutant-general, Ian Hamilton.

Roberts was riled—both by the implication that the army collectively was against conscription and because he saw Hamilton as one of his own protégés. In 1911 he replied with *Fallacies and Facts: An Answer to 'Compulsory Service'*. He drew attention to the point that it was a Liberal secretary of state for war who was responsible for breaching the principle of silence which had cloaked the views of military members of the Army Council since its inception. And he then went on to speculate as to why Haldane's choice of author had alighted on Hamilton rather than on Sir William Nicholson, the chief of the imperial general staff. He implied that Hamilton's opposition to conscription was the exception rather than the rule among senior officers.[41]

Roberts's naming of Nicholson was shrewd: the chief of the imperial general staff did indeed favour conscription. However, the senior serving officer who publicly declared his support for the League was not Nicholson but Major-General Henry Wilson, the commandant of the Staff College. In 1909 Wilson gave two lectures on the subject to his students.[42] In August 1910 Wilson became director of military operations. It was in his hands that Anglo-French staff talks assumed both continuity and urgency, and it was in him that the idea of a firm military commitment to Europe found its most forceful exponent in the years before the First World War. Conscription for Wilson was not simply a matter of home defence, but also a reflection of his belief that Britain needed an army structured on Continental principles.

Wilson was a member of 'the Roberts ring'. He had acted as Roberts's private secretary when Roberts was commander-in-chief, and he continued to refer to him as 'the chief'. Roberts and Wilson had a further link: they were Anglo-Irish. Both were Unionists, opposed to home rule for Ireland, and opposed to the coercion of loyalists within Ulster.

In 1893 a third Anglo-Irishman, Garnet Wolseley, had written to the duke of Cambridge, on the occasion of the second home rule bill:

If ever our troops are brought into collision with the loyalists of Ulster and blood is shed, it will shake the whole foundations upon which our Army rests to such an extent that I feel our Army will never be the same again. Many officers will resign to join Ulster and there will be such a host of retired officers in the Ulster ranks that men who would stand by the Government, no matter what it did, will be worse than half-hearted in all they do. No Army could stand such a strain upon it.[43]

It was in Ireland that the security of the empire and the army's politicization through its imperial experience came together. It was Ireland that set at odds the professional self-regard of the army and the principles of the Liberal government. The Curragh incident of 1914 therefore stands as the logical conclusion to a century of continuous influences. It was not, as Robert Blake would have it, 'an exception to all normal rules'[44]—the normal rule being that the army was apolitical. It was instead the single event that illustrates most starkly the political consequences of professionalism.

In the short term the indications of possible difficulties in using the army in support of Irish home rule, particularly if there was loyalist opposition to the coercion of Ulster, were sufficiently plentiful to cause the king to raise his concerns with Asquith, the prime minister, in the summer of 1913. In that year also, Henry Wilson established contacts with the Conservative leader, Bonar Law, and the Unionists became sufficiently aware of the army's position to realize the advantages of developing and exploiting it. In Ulster itself, Lord Roberts had taken a key role in the establishment of the Ulster Volunteer Force from 1912. He put forward the name of its commanding officer, Lieutenant-General Sir George Richardson, who had served under him in India. The army's close link with the Ulster Volunteers was confirmed by the fact that 62 per cent of the latter's divisional, regimental, and battalion commands were held by former officers.[45] Roberts publicly congratulated Edward Carson, the Ulster Unionist leader, on the success of the Larne gunrunners. Both his stance, and the existence and composition of the Ulster Volunteers, made it quite clear that the army would not be a compliant tool in any policy to coerce Ulster into home rule. The Curragh incident was not a bolt from the blue: the battle lines were being drawn up in advance.

The events of March 1914 are relatively straightforward.[46] Given the formation by both sides of volunteer forces, the government was worried about the security of its armouries. It therefore instructed the general officer commanding in Ireland, Sir Arthur Paget, to increase the guards mounted over them. Paget felt that such action would be provocative and even counter-productive; although he did not have access to the same intelligence as the government, he believed that the existing guards were sufficient. Paget was therefore summoned to the War Office for talks on 18 and 19 March.

The immediate object of the meeting was to impress upon Paget the need to reinforce the guards. There was no question at this stage of coercing Ulster. None the less, the opportunity was taken to discuss possible courses of action if the need for coercion did arise. J. E. B. Seely, the secretary of state for war, made clear that officers could not resign if they refused to obey orders when asked to act in support of the civil power. They would simply be dismissed the service. For career officers, committed not only by a sense of vocation but also by the progressive acquisition of pension rights, the implications would be catastrophic. Seely did, however, suggest that some indulgence might be shown towards officers who were domiciled in Ulster.

Significantly, Paget was given no written instructions. On his return to Dublin on 20 March, he summoned his senior officers to a conference. Paget was neither particularly intelligent nor particularly subtle. He used strong language. He suggested that violence in Ulster was imminent, and that active operations in the north were envisaged. He then gave his officers an ultimatum. Those domiciled in Ulster could disappear until things were over. Those not so qualified (and they included those resident elsewhere in Ireland) who were not willing to serve could tender their resignations, but the resignations would not be accepted and instead they would be dismissed the service.

Paget had therefore set up a hypothetical situation, and offered his officers a choice as to how to respond. Many of those who reacted against his ultimatum protested that, if in the event of trouble breaking out they had been given a direct order to aid the civil power, they would have obeyed. The officers of the 3rd Cavalry Brigade, commanded by Hubert Gough, were particularly incensed by the ultimatum. In Gough's own regiment, the 16th Lancers, all sixteen officers requested dismissal. In the other two regiments of the brigade, the 4th Hussars and 5th Lancers, seventeen out of nineteen and seventeen out of twenty opted for dismissal.[47]

Gough was summoned to London, together with the commanding officers of the 5th Lancers and the 16th Lancers. The cabinet, anxious to close the matter, approved a memorandum for submission to the recalcitrant cavalrymen. It did no more than assert that the incident was the product of a misunderstanding; it then emphasized that it was the duty of all soldiers to obey lawful commands. It was therefore a statement of constitutional norms. But it did not address the issue of Ulster, and it was insufficiently specific to appease Gough. Seely and the chief of the imperial general staff, Sir John French, therefore added two paragraphs, which effectively committed the government not to use coercion. This was the guarantee which Gough wanted. He promptly accepted it, adding as a gloss his own interpretation. When the cabinet discovered what had happened it disowned the guarantee in the form in which Gough had negotiated it, and both Seely and French were compelled to resign.

The key question at the heart of the Curragh incident is whether it was the product of accident and confusion, or of conspiracy and collusion. The principal player in setting events off on a course so different from that apparently anticipated in the War Office meetings of 18 and 19 March was Sir Arthur Paget. Why were his actions in Dublin on 20 March so at variance with the instructions which he had been given on the previous day? Paget was certainly hotheaded, impulsive, and unreflective: such attributes support the case for confusion. But was he really as stupid as this interpretation would suggest? The account of the meetings in the War Office given by the adjutant-general, Lieutenant-General Sir J. S. Ewart, are straightforward enough: they do not suggest any ambiguities sufficient to permit Paget to draw his own conclusions. It behoves us, therefore, to consider the suggestion that there was a conspiracy. Specifically, was Paget got at by Wilson and Roberts, and possibly by others?

Henry Wilson's role in the crisis is well documented. As director of military operations in the War Office, he was situated at the hub of affairs. He kept both the press and the Unionist leaders abreast of events. He also saw Gough before his meetings in London, and took the opportunity to bolster his resolve. Wilson had emerged from his first stint at the War Office—in Roberts's time—convinced that civilian government of the army was 'vicious in theory and hopeless in practice'.[48] By 1914 he seems to have come close to putting constitutional principles in reverse, suggesting not merely that the army should run its own affairs, but that it should also run those of the country as a whole. A. P. Wavell, who was on Wilson's staff, wrote to tell his father that, when the news of Gough's guarantee broke, Wilson had triumphantly announced: 'The Army have done what the Opposition failed to do', and that it 'will probably cause the fall of the present Government'.[49]

Roberts's role is more shadowy, although Asquith for one was convinced that the aged field marshal was at the centre of the whole affair. Certainly Roberts was no friend to the Liberal government. He was an advocate of conscription and an opponent of Haldane's reforms; he had aligned himself with the 'Diehards' in opposing the reform of the House of Lords. Lord Milner, his ally on that occasion, reported in January 1914 that Roberts was 'in a great state about the effect of the whole thing on the Army'.[50]

The only direct evidence of Roberts's attitude during the crisis is a telephone conversation with Sir John French on 21 March, as reported by French. Both Roberts and Wilson feared that the chief of the imperial general staff, known as a Liberal for his pro-Haldane views, would not stand up to the government. According to French, Roberts said, 'I hope you are not going to associate yourself with this band of (certain epithets were used which I could hardly catch). If you do you will cover yourself with infamy.' French replied: 'I must do my duty as a soldier like everyone else and put up with whatever consequences may ensue.' Roberts then said, 'goodbye', and rang off. The field marshal subsequently denied that the epithets which he used to describe the government and which Sir John tactfully could not catch were 'swines and robbers'.[51]

Once Gough got the guarantee, Roberts's position became more overt. He was clear that the army and the Unionists must cleave to Gough. When the cabinet pointed out that the guarantee went beyond the wording it had agreed, Roberts—and Wilson—were part of a pressure group pushing French to resign.

What is extraordinary about the behaviour of Roberts and Wilson in the Curragh crisis is that they were willing to split the army in their pursuit of Unionism—that they were prepared to give their political objectives priority over their sense of professional solidarity. In one sense of course there was nothing new in this. The divisions within the army between the duke of Cambridge and Wolseley, and between Wolseley and Roberts, were the natural consequences of ambitious officers exploiting opportunities so as to get political support, and the civilian politicians themselves had developed the divisions which had followed.

In 1914, however, those in the army who opposed Gough's stand did so precisely on the grounds that the army must remain united. This was the other side of professionalism. If the army—as Napier and Wolseley had both argued—was the embodiment of the nation, then it must be above party. The man who did most to contain the fall-out from Paget's remarks on 20 March was Sir Charles Fergusson, the general officer commanding the 5th Division. After Paget's presentation of the 'ultimatum', Fergusson toured the units of his command. He 'spoke of the far-reaching consequences of a disruption of the Army. That the country without a disciplined and united Army would be at the mercy of the mob. I alluded to the probability of even bigger questions arising if the Army broke up, that the Monarchy, Society and the Empire itself might be shattered.'[52]

However, Fergusson's reference to the empire is also the key to the positions of Roberts and Wilson. Obviously neither of the latter two wished actually to split the army: they hoped to carry the whole army with them. In their eyes what justified the risk to the unity of the service was the greater risk to the empire as a whole presented by Irish home rule and by the coercion of Ulster loyalists. Roberts told Asquith that civil war over Ireland would not be confined to Ulster, nor to Ireland, nor to Great Britain. Its effects in India could, he said, prove incalculable. It could also provide the opportunity for a European power to take advantage of Britain's domestic difficulties.[53] Therefore for both divisions in the army—those who did support Gough as well as those who did not—the empire and its maintenance were crucial factors in determining their political activity.

The Curragh incident was therefore the denouement to the army's politicization in the course of the nineteenth century through its involvement in the empire. Imperial soldiering was the basis on which ambitious officers built their careers. The context in which they developed their political skills and convictions was consequently professional. But when the paladins of empire came home to high office in London, they applied those attributes with effects that were more obviously political.

Moreover, the Curragh was not just a culmination. Its significance lay also as a pointer to the future. It emphasized that the British army entered the First World War deeply politicized and well versed in the arts of political intrigue. It highlights the fact that, for all the good relations between some soldiers and some Liberals (especially Haldane), the more general distrust between the army and the Liberals evident since the 1880s had persisted and even intensified. Its consequences were to be played out against the background of a war that was primarily a war for Europe, but was also of course a war for the defence of the empire.

6

The Politics of Integrated Control 1895–1918

When Clode wrote *The Military Forces of the Crown* in 1869 he was penning a plea for dual control. What worried him was the consolidation of military administration in 1854–5. With the removal of the commissariat from the aegis of the Treasury, he saw supply officers with financial responsibilities being given commissions in the army and thus subordinated to the commander-in-chief and hence to the crown. Furthermore, the flow went in both directions. Soldiers moved into the War Office and performed tasks that before 1854 would have been carried out by civilians. He conjured up an image of an ever-expanding army; he feared that the establishment of the army's own supporting services, staffed and administered by soldiers—of transport, medical care, education, religion, and supply—would create a separate and self-contained military caste.[1]

Clode's vision was somewhat hysterical. He exaggerated the threat of military domination after 1855 just as surely as he distorted the reality of dual control before 1855. But in condemning Clode, care must be taken not to reject entirely the burden of his points. Dual control, ironically enough, was as strong in his time as at any point in the history of the army. What followed it, beginning in 1895 with the retirement of the duke of Cambridge and the partial implementation of the recommendations of the Hartington report, was integrated control. The growth of the War Office, and the incorporation within it first of the commander-in-chief and then of the chief of the general staff, were manifestations of this process.

In reviewing the evolution of Britain's machinery for the administration of defence over the last century, it is easy to fall victim to a Whiggism as misleading as that brought to bear on the 1689 settlement and the notion of dual control. There is a tendency to see the steps from Lord Hartington to Michael Heseltine as those of the onward march of civilian control—of the subordination of the army to the domination of the constitution. Moreover, this is a perspective that owes its origins not just to the Whig view of history. It has found confirmation in the re-education of liberal historians in the nostrums derived from Clausewitz: it is evidence of the truism that war is an instrument of politics.

Clode was more right than these perspectives allow. The establishment of integrated control did not result in the political emasculation of the army. Rather, it expanded the soldier's field of influence. In practice, if the army was subordinated to anything, it was to ministerial direction rather than to the constitution. But by moving its command headquarters from the Horse Guards to the War Office, the army increased rather than diminished its leverage over the secretary of state for war. Co-operation could replace confrontation. Integration allowed the secretary of state for war to hear and consult military experts; it encouraged him to become the advocate of military efficiency rather than the spokesman of financial stringency or the defender of constitutional norms. Clode's argument—that the administration of the army was militarized rather than civilianized in the wake of the restructuring of 1854–5—may be overdrawn. But it provides an important corrective to those who put the opposite interpretation. In a sense both sides—military as well as civilian—gained in political power: fusion rather than fission enabled them to maximize its potential.

What particularly irked Clode as he wrote was the 1867 report of the Strathnairn committee on field transport. Until 1854 transport had been in the hands of the commissariat, a civilian body administered by the Treasury. So fundamental was the shift in attitudes by 1867 that the Strathnairn committee was made up entirely of the army's employees, and it addressed the issue exclusively as a military problem. Constitutional considerations did not form part of its deliberations. Its recommendations for the reform of the army's transport and supply services included the proposal that the comptroller of the army's chest in stations overseas be subordinated to the local general officer commanding: in other words the principle that the command of the army should be kept separate from its

financing was to be breached. One member of the cabinet, according to Clode, described the arrangements as 'little short of Revolutionary'. And yet when the Commons came to discuss the report,
attendance in the House dwindled at one stage to thirteen members, of whom nine were soldiers.[2]

Clode was right in his forecasts. In 1887 a royal commission,
chaired by a judge, Sir James Fitzjames Stephen, tackled the question of the army's 'warlike stores'. The immediate issue was the
failure of weapons issued to troops in the Egyptian campaigns of
1882 and 1884–5. But the commission refused to observe instructions that they accord their brief a narrow interpretation. Wolseley,
as the commander-in-chief on both occasions, was a prime witness:
characteristically he expressed the hope that the secretary of state
for war should be a man of military experience. Something of
Wolseley's thinking was to be found in the commissioners' conclusions. They criticized civilian controls in terms characteristic of
soldiers themselves. In their view, the principal difficulty was that
the secretary of state for war was in office for too short a time, and
was too susceptible to the influences of party politics. They concluded that the effects of civilian control were:

First. It has no definite object.
Secondly. It has no efficient head.
Thirdly. It has no properly organized method of dealing with technical
questions which arise as to the construction or purchase of warlike stores.[3]

The Stephen committee recommended that soldiers should head
the supply departments and that those soldiers should in consequence be directly answerable to parliament.   ·

Edward Stanhope embraced many of the suggestions made by
Stephen's committee. The commander-in-chief was given charge of
all the principal departments of the army, except finance and
manufacture, and supply and transport were made directly answerable to the quartermaster-general. But Stanhope did not go so far
as making soldiers their own spokesmen in parliament. The army
therefore argued that it had been given accountability without
power. Parliament could still refuse money, so weakening the
army, while the army itself remained muzzled. Neither the duke

of Cambridge nor Lord Wolseley would accept responsibility on these terms.

The upshot was the royal commission chaired by Lord Hartington in 1890. The results of its deliberations have been briefly considered in Chapter 3. The abolition of the post of commander-in-chief was to be the precursor of a three-tiered revision in the administration of the army. First, a defence committee, formed of soldiers, sailors, and ministers, should be charged with the consideration of the defence estimates before they were submitted to the cabinet; in particular it should be responsible for relating the estimates to an overall plan of imperial defence. Secondly, a War Office council should replace the commander-in-chief. Thirdly, a general staff should be set up, with the task of preparing plans for the defence of the empire and so guiding the defence committee in its deliberations on the army estimates. The essential point in the present context was that in both the defence committee and the War Office council soldiers and civilians would serve alongside each other: the recommendations which they made would therefore constitute joint opinions.

The Hartington commission's proposals for integrated control were entirely in step with the reforming orthodoxy of the day. In parliament Sir Charles Dilke, Lord Charles Beresford, Lord Wolseley, and—especially—Colonel John Colomb were calling for measures on similar lines. In 1892 Dilke wrote a book in conjunction with Spenser Wilkinson. Called *Imperial Defence*, one of its principal themes was the need for integrated command. Its authors advocated the appointment of a commander-in-chief or chief of the general staff with the task of advising 'the Cabinet upon the conduct of possible wars, on the understanding that he will be called upon to execute his plans if they are accepted'.[4] The means by which they planned to achieve integration were therefore closer to the arrangements pertaining in Germany than to the structures adumbrated by the Hartington commission, but the objective was the same.

Despite this groundswell of support, the implementation of the commission's recommendations was not immediate. Only the War Office council was established in 1890, and as its agenda was determined solely by the secretary of state its contribution to the principle of integration was limited. It required the Boer war and the subsequent report of the Esher committee to complete the process initiated by Hartington. The Committee of Imperial Defence—a descendant of a cabinet committee on defence created in 1894–5—was formally set up by Arthur Balfour in 1902. The office of commander-in-chief was abolished in 1904—although its powers had of course been clipped in 1895. The Army Council replaced the War Office council at the same time, and the general staff was established in 1906.

Even now the level of integration was not as complete as its advocates on the Esher committee would have liked. The members of the Committee of Imperial Defence, whether military or civilian, were equals. But the committee was an advisory body only, and the execution of its recommendations was the responsibility of individual ministers. The Army Council was a joint civil–military body, but it was answerable to the secretary of state for war, and he alone was responsible to the cabinet. Significantly, the most powerful agency for the expression of army opinion, the general staff, was the last of the new bodies to be created.

What gave reality to the idea of co-ordination in defence was less its organization than the attitudes of the personalities involved in its business. The appointment of Haldane as secretary of state in 1906 was vital in bedding down the revised arrangements. The parallel with another great reforming Liberal war minister is obvious: Cardwell achieved so much because he had the support of a great many officers, including of course Wolseley. Haldane was the same. He won the respect and friendship of the army's rising stars, most notably Douglas Haig. Even in 1939, according to Sir John Kennedy, Haldane continued to be the yardstick by which soldiers in the War Office judged the worth of his successors in office.[5]

Haldane used and developed both the Committee of Imperial Defence and the Army Council. Although ultimate responsibility remained with the minister or ministers, the members of the two committees—whether soldier or civilian—were treated as equals. Discussion was freer and franker, and Treasury intervention effectively excluded at these levels. Thus the agencies of military–political co-ordination gained in authority, flourishing because of the concerns generated by the South African war, by the consequent movement for 'national efficiency', and—in due course—by the threat from Germany. The perception of danger meant that the advice of the soldier—or sailor—was trusted more often than not.

Although used to advantage by Haldane, the Committee of Imperial Defence was of course not in any formal sense a fiefdom of the War Office. It was shaped by Balfour to be a tool of the prime minister. He was the only politician to sit on it ex officio. But neither of his Liberal successors—Campbell-Bannerman and Asquith—showed the same interest in its deliberations as he had done. Thus, after 1905, it derived its continuity from its secretariat. Both this and the subcommittees which the main committee spawned were dominated by professional advisers. The Committee of Imperial Defence was a progenitor of modern cabinet government, and yet the key players in its maturation were neither politicians nor civil servants, but soldiers and sailors.[6]

Pre-eminent in this group—and a personality as important as or even more important than Haldane in establishing the principles of integration and co-ordination—was Maurice Hankey. Hankey was naval assistant secretary of the Committee of Imperial Defence between 1908 and 1912, its secretary from 1912 until 1938, and secretary of the war cabinet from 1916 until 1919. He established the procedures that shaped cabinet government: the setting of agendas, the taking of minutes, the proliferation of specialist committees, and the centrality of the prime minister to the processes of national security. But Hankey began life as a Royal Marine. Service pedigrees persisted after Hankey. In 1938 his successor as secretary of the Committee of Imperial Defence, H. L. Ismay, was a soldier. By then, Ismay reckoned 95 per cent of the business of the Committee of Imperial Defence was decided without reference to the full cabinet.[7] Service advice and expert opinion were every bit as pervasive as Clode's apocalyptic visions had forecast.

Neither Hankey nor Ismay was a faceless bureaucrat, devoid of his own ideas. Hankey in particular had cogent views on the strategy appropriate to the conduct of the First World War. But both exercised their political influence through emollience, as facilitators rather than as blusterers. Their power, although unsanctioned by election and launched by means of military careers, remained uncontroversial because of the manner in which they chose to exercise it.

Very different was the penchant for intrigue and lobbying which was the legacy of the army's imperial experience. The soldier who most obviously acted as the bridge between the politics of empire and the politics of integrated control—and arguably the most notoriously 'political' of all British army officers—was Henry Wilson. His military career, begun under the tutelage of Roberts, reached its climax with his appointment as chief of the imperial general staff in 1918. He brought the in-fighting skills of his mentor to bear on the problems of grand strategy and coalition warfare. One of Wilson's military colleagues said that even seeing a politician could induce in him 'a state of sexual excitement'.[8]

Wilson was commandant of the Staff College between 1907 and 1910. In 1908 he was criticized for encouraging his students to debate 'high political points' on the grounds that these were outside the competence of the soldier. Sir Henry Rawlinson, who had preceded Wilson at the Staff College, said that the practice pre-dated Wilson's arrival at Camberley. Colonel L. E. Kiggell, who became commandant in 1913, observed that, although a distinction could be made between 'the need to study politics in relation to strategy' and the 'need to study one's own political problems', 'politics were at the back of all strategical problems' and thus no clear division could be maintained in practice. Significantly, the most cautious view expressed in this debate emanated from William Robertson, Wilson's immediate successor at the Staff College, who agreed that policy determined strategy, but also stressed that the setting of policy was not the task of a general staff: its job was to advise on the methods by which policy might be put into effect.[9]

The issues which Wilson was accused of encouraging his students to debate were those of foreign affairs. Formally Britain was not committed to a military alliance with France. To that extent any discussion resting on the presupposition that it was could be deemed controversial. But Wilson—albeit in a more provocative fashion—was thereby pursuing the goal of fusion as fully as was the Committee of Imperial Defence. The education of staff officers in an era of integrated control demanded that consideration be given to the problems on the cusp of civil and military responsibilities.

What was regarded as a debatable practice in 1908 would seem an entirely proper activity in the Staff College of today. In 1958

Brigadier C. N. Barclay wrote that 'the able officer must assert himself as a co-partner in the business of strategy'.[10] The International Institute for Strategic Studies was formed in the previous year, 1957; the venerable Royal United Services Institute added the suffix 'for Defence Studies' in 1971. Since the 1960s much professional military education has been political in the sense in which the word was being used at the Staff College in 1908. Both the existence of strategic studies and the fact that their pursuit constitutes a legitimate activity for soldiers provide evidence of integration.

However, co-ordination is not the same as agreement. Patently, it would be absurd to claim that the creation of the Committee of Imperial Defence, the Army Council, and the general staff ushered in a decade of civil–military tranquillity before the outbreak of the First World War. Descriptions of the War Office prior to Haldane's arrival dwell not on the harmony created by the new structures, but on divisions and discord. After his departure, in 1912, the old tensions were restoked by Roberts, Wilson, and others, and found in the Curragh a climax consonant with all that had gone before.

The Committee of Imperial Defence did not have either the bottom or the remit to cope with these sorts of problems. It achieved a great deal in preparing Britain for mobilization; it resolved issues that affected both the army and the navy if and when both these services were agreed that they wanted a resolution. Otherwise it dealt with issues that were the responsibility of neither—that would have fallen into the gap between the two.[11] It was too weak to achieve more. For the army the general staff was the prime planning body. The navy, under the professional direction of Jackie Fisher as first sea lord, resisted the Committee of Imperial Defence: it used it more after 1911, but as a counterweight to the army's general staff (since the navy did not possess an effective staff) rather than as a forum for joint discussion.[12] Thus the individual services, not the Committee of Imperial Defence, made plans. Only once before 1914, on 23 August 1911, during the second Moroccan crisis, did it look at Britain's overall strategy in the event of war, and even then it did not make strategy—it simply opted for the plans of one service (the army) rather than those of the other (the navy).

Most important of all, the Committee of Imperial Defence was essentially an organization for peacetime pontifications. It had no executive powers. On the outbreak of war, the cabinet resumed its primacy in policy. The committee itself languished. Although the defenders of the Committee of Imperial Defence would claim that it influenced and shaped the subsequent machinery for the direction of war,[13] in 1914 itself Britain possessed no such machinery.

Asquith's solution was to appoint Kitchener secretary of state for war. At one level this is striking evidence of integration. Kitchener was the country's most distinguished soldier and yet the government—and a Liberal government at that—appointed him to be the army's political head. Kitchener's elevation also serves to reinforce another point—the way in which imperial proconsuls learnt their political skills in the empire and then returned to exercise those skills at home. Kitchener's capacity for intrigue had already been made abundantly clear by his tenure of the chief command in India. His guile had been employed in order to amass power in his own hands: Curzon's criticisms had, at least initially, rested on the grounds of practicality rather than of constitutionalism. Kitchener's mania for centralization was equally evident in Britain. As secretary of state Kitchener effectively combined three roles in one—that of his declared office, that of chief of staff to the secretary of state, and that of chief of the imperial general staff. He had left—as the events of 1914 showed—the potential for chaos in Delhi; in the same year he created chaos in London.

Between 1906 and 1914 the general staff had evolved as the political focus for the army's professional concerns. It was through the general staff that its military preparations for war on the Continent took shape, and it was the general staff and its representatives that gave the army a voice in strategy both in the 1911 Moroccan crisis and on the occasion of the outbreak of war in 1914.

But in 1914–15 the general staff was shattered. The Curragh set the director of military operations (Henry Wilson) against the chief of the imperial general staff (Sir John French). French resigned, and with him went a sense of continuity. The mobilization of the army then drew many of the general staff's key players to staff appointments in France, and away from London. Kitchener's love of centralization and of independent control stood at odds with the idea of collective responsibility inherent in a general staff: efforts to revive it would fail while its functions remained unclear. Hankey and the secretariat of the Committee of Imperial Defence stepped into the breach. But this had the effect of excluding the professional heads of the services from the formulation of strategy, since the latter was now discussed not in the Committee of Imperial Defence but in the cabinet, a body on which neither the chief of the imperial general staff nor the first sea lord sat.[14] Thus the army lost its political voice in 1914—despite the fact that, paradoxically, its most senior active officer was elevated to high political office.

Strategy was therefore settled by the ministers, principally of course by Kitchener himself, but also by the first lord of the admiralty, Winston Churchill, who similarly acted as his own head of operations and chief of staff. Cabinet government itself broke down. Strategic decision-making became confused and contradictory.

This in itself was a situation in which the political instincts of the army would probably have asserted themselves. But there was a further problem—the declining legitimacy of the government in party political terms. An election was due in 1915, and it was one which the Liberals could not expect to win. The party leaderships of both the Liberals and the Conservatives had to contend with recalcitrant backbenchers. A minority of Liberals had opposed Britain's entry to the war and were correspondingly disenchanted with Asquith. The Conservatives were frustrated by their leaders' passive collusion with Asquith's government for the sake of the national interest; they preferred to enjoy the fruits of office.

The crisis of May 1915, which led to the fall of Britain's last Liberal government and to the creation of a coalition, cannot therefore be attributed simply to the intervention of the army in politics. Some historians have seen it as a conspiracy engineered on the front benches—even by Asquith himself—in order to bring the backbenchers to heel. From a Liberal perspective, the coalition postponed the need for an election and so sustained the Liberals in office. Others would argue that the clash at the Admiralty between Churchill and Fisher, which resulted in Fisher's resignation, was more influential. Fisher's actions are—parenthetically—illustrative of the navy's own well-developed political instincts; in particular he exploited the Conservatives' profound distrust of

Churchill to shatter the truce on which the continuation of Liberal government rested.[15]

In recognizing the importance of these points, however, we must not go to the opposite extreme and lose sight of the army's role in the events of May 1915. They are particularly instructive because of what they tell us about the army's political instincts. Sir John French, who had been recalled in August 1914 to be commander-in-chief of the British Expeditionary Force, was not suddenly 'politicized'. Senior generals are inherently predisposed to behaviour that can be construed as political precisely because their jobs employ them at the confluence of military and political power. In 1915 French was motivated by professional frustration with Kitchener. Professionalism led to political intervention. The opportunity for intervention arose as a result of the declining legitimacy of the political order which had hitherto circumscribed the army.

French's anger with Kitchener found its immediate origins on 1 September 1914. The German advance through Belgium in the opening weeks of the war had hit the British army with particular ferocity. French's nerve was shaken. As the BEF fell back southwards in conformity with the general allied retreat, he had to weigh two obligations—the need to support the armies of France and the requirement to preserve intact Britain's only army. He opted to elevate the latter over the former, announcing his intention to leave the line and fall back behind the Seine to regroup and recover. Kitchener, who at the outset of the war had impressed on French the very duty to which he was now responding, donned his field marshal's uniform and crossed over to France. He overruled Sir John. In doing so, he not only stressed the obligation to France, he also implicitly reinforced the doubts about French's moral courage. Although technically Kitchener was French's political superior, he had behaved as his military master.

Two months later, at the height of the first battle of Ypres, doubts about French's resolution were again voiced—this time to Kitchener by France's commander-in-chief, Joffre, and his deputy, Foch. Kitchener promptly offered to remove French and replace him with Sir Ian Hamilton. Joffre and Foch assured Kitchener that that would not be necessary, particularly in the middle of a battle. But Foch told his friend Henry Wilson, now employed as assistant chief of staff at the headquarters of the British Expeditionary Force, what had occurred. With his customary eagerness for such gossip, Wilson passed on the intelligence to French. French was apoplectic with rage, and dispatched his aide-de-camp, Captain F. E. Guest, significantly a Liberal MP, to protest to Asquith.

During the winter of 1914–15, the personal antagonism between the two became clothed in divergent strategies. French could not accept Kitchener's conclusion that the deadlock on the western front would continue in 1915. Kitchener's preference was to husband Britain's manpower, building up the New Armies for a major effort in 1917.[16] French wanted more men immediately, arguing not unreasonably that large armies would be of no value in 1917 if the war were already lost by then. Not only that, French also reckoned that he could break the defensive systems of the western front if he had sufficient shell to blast a hole through the line. This was a common cry from all field commanders in early 1915 and the fact that the BEF did not get as many shells as it demanded was neither an experience unique to it nor entirely the fault of the War Office. The problems in shell supply arose from the more general difficulties of industrial mobilization. But Kitchener's combination of excessive centralization and obsessive secrecy meant that the blame ascribed to the War Office both by Lloyd George (then chancellor of the exchequer) and the press was not easily shifted. Certainly it was grist to French's mill.

Sir John made sure that his views were known in London. On 3 January 1915 he wrote to Kitchener, opposing the opening of new theatres of operations: both the Dardanelles and Salonika were under consideration. He copied his letter to Asquith and to Churchill, and it was then circulated to the entire cabinet. Kitchener was furious. He had been effectively circumvented in his self-appointed capacity as the cabinet's professional military adviser. Ten days later, French made his own case to the War Council, a subcommittee of the cabinet.[17]

Although Kitchener's national reputation was still intact, many of his cabinet colleagues found his insensitivity to the norms of civil government a source of increasing frustration. Kitchener's colleagues responded therefore to the evidence of Sir John French's frustration. But they could not dispense with Kitchener's services while his massive frame and piercing eyes remained the symbol of national determination. The removal of Kitchener might shatter the Conservative forbearance on which Liberal survival rested.

However, the conversion of the Tory press to an anti-Kitchener position was already in train. Henry Wilson hated Kitchener. Wilson was encouraged by Sir John French to believe that he would have liked to appoint Wilson as his chief of staff rather than as the latter's deputy, but that the appointment had been blocked by both Asquith and Kitchener. This was hardly surprising. After all, Wilson had crossed the prime minister over the Curragh, and he was now ridiculing Kitchener over the creation of the New Armies and the strategy of delaying their employment until 1917. Wilson cultivated H. A. Gwynne, a former war correspondent and now the editor of Lady Bathurst's Unionist paper the *Morning Post*.[18]

The newspapers owned by Lord Northcliffe had also turned against Kitchener. Northcliffe was frustrated by the restrictions imposed on his reporters by the War Office. The *Daily Mail* in particular blamed Kitchener for the shells crisis. Repington, *The Times*'s correspondent, had become alienated from his former chief, and had been taken in by French as his personal guest at general headquarters.[19] On 1 May 1915, Northcliffe invited French to make 'a short and very vigorous statement' which would 'render the Government's position impossible'.[20]

A week later French was provided with his opportunity. On 9 May a British offensive at Aubers ridge failed, not least because of artillery problems. In the previous month French had reassured Kitchener about shell supply, and his statements had provided the basis for a complacent speech on the subject by Asquith in Newcastle on 20 April. But now French opted to divert the blame for the defeat at Aubers ridge from his own shoulders to those of Kitchener, using the lack of high explosive shell as the explanation for the setback. His earlier assurances, he pointed out, had related only to the supply of shrapnel. On 14 May 1915, *The Times* carried an article by Repington blaming the government for the shells crisis, and attacking both Asquith and Kitchener. Conservative backbenchers were outraged, and their leader, Bonar Law, had to follow them in their fury. In an independent move, Fisher resigned as first sea lord the following day. On 17 May Asquith called for a coalition government.

Britain's last Liberal government did not fall on the say-so of the army alone; it is probably even true that the army's involvement was peripheral to the main political crisis. But the point still remains that Sir John French was conspiring to bring down Kitchener at the very least, and that if successful the effects of his actions were bound—given the popular prestige which Kitchener enjoyed and given the Liberal government's reliance on him for any semblance of strategic authority—to shake Asquith himself. Furthermore, French was not simply the passive agent of press agitation or of a front-bench conspiracy.[21] He took an active part in the whole process, to the point of sending two of his officers, Brinsley Fitzgerald and F. E. Guest, to London to brief Lloyd George and the Conservative leaders, Arthur Balfour and Bonar Law.[22] After the crisis, Fisher said to Balfour: 'It was better to wreck the Government than the Navy.'[23] Such sentiments, appropriately rephrased to embrace the army, might well have been uttered by French. After the war, in his memoirs, French was quite happy to claim the responsibility for the fall of the government and to assert that he had acted in the national interest.[24]

There is a tendency when considering civil–military relations in the First World War to concentrate attention on Lloyd George and his relationship with the generals. But in the process the implications of May 1915 for the history of civil–military relations tend to be neglected. They must not be. Here is a case when the army was implicated in the fall of a government. Furthermore, it was a crisis which formed the denouement to thirty years of increasing tension between the army and the Liberals.

After May 1915 the principal focus of discord in civil–military relations shifted. French was replaced by Douglas Haig in December 1915. Haig was as distrustful of Kitchener as his predecessor had been. He was, after all, Haldane's man, and Kitchener had not followed Haldane's principles for expansion when he had set about the creation of the New Armies. But Haig's sources of political power were less personalized, less fragile, and ultimately less fractious than those of his predecessor. He had kept his hands clean during the Curragh crisis. He did not trust Henry Wilson, and so

did not get drawn into his intrigues. Although H. A. Gwynne wrote
to Haig in unctuous terms on the occasion of his appointment as
commander-in-chief, inviting him to side with the press against the
politicians, Haig kept his distance, finding Gwynne conceited and
self-satisfied.[25]

Haig's props were more institutional. First, as already discussed
in Chapter 3, came the king. Second, and even more important,
was the general staff, revitalized with the appointment of Sir William Robertson as its chief, also in December 1915. From June 1915
onwards Haig and Robertson lobbied the king and his advisers on
the need to revive the general staff in order to clip Kitchener's
powers. The idea that the chief of the imperial general staff should
be responsible only to the War Committee (successor to the War
Council) and not to the secretary of state for war was adopted first
by George V, and only subsequently approved by Asquith. It
formed the condition on which Robertson accepted office, and it
made him rather than the secretary of war the government's principal strategic adviser. Its effect was to shift the debates about the
limits of professional responsibility away from general headquarters
in France and back to London.[26]

However, Robertson's political leverage depended on more than
the formal arrangements governing the general staff. It also rested
on the fact that there had still been no election. Parliament had
become increasingly redundant in political debate—or certainly in
so far as it affected strategy—and the main critics of the government were to be found not at Westminster but in the newspapers.
Both *The Times* and the *Morning Post* were firmly supportive of the
generals. The failure of the Gallipoli campaign in 1915 and the
surrender of Kut in Mesopotamia in April 1916, reinforced by the
royal commissions appointed to investigate both set-backs, discredited the politicians' direction of strategy and correspondingly
enhanced the status of the professional experts.

Robertson was therefore well placed to develop the political
power of the soldier. But he was not politicized in the sense that
Wilson was. Indeed, he disliked Wilson, and, like Haig, he had
kept his distance during the Curragh. At the Staff College before
the war his had been the voice of caution concerning the political
education of its students. During that time he wrote on an essay

by A. P. Wavell: 'The discussion of questions of policy and political matters generally leads to no practical result, nor benefit of any kind to the soldier, nor is it his business.'[27] He emphasized both then and again in the summer of 1915 the subordination of war to politics and the need for those responsible for the shaping of strategy to conform to political objectives.

This does not mean that Robertson was a political neuter. He was a professional soldier. He was perhaps the most obvious embodiment of that phrase in the entire history of the British army. A former domestic servant, he had enlisted as a private and he eventually made his way to the top of the tree by becoming a field marshal. His views on strategy, particularly now that they can be examined in isolation from the invective of Lloyd George's memoirs, are characterized by common sense, pragmatism, and shrewdness. What gave Robertson political significance, and even fostered in Robertson views on politics, was his professional commitment rather than any political ambition *per se*.

Britain was engaged in a world war. Its conduct was the activity to which all other national preoccupations were subordinated. These were the factors that willy-nilly pushed the general staff into the political limelight. Robertson was the country's principal strategic adviser. His actions were determined by the pursuit of strategic effectiveness, not of political effect. His advice had even greater consequences than might otherwise have been the case for two reasons. First, the army's institutions were—at least until the creation of a war cabinet in December 1916—more coherent than those set above them for the higher direction of the war. Secondly, the coalition government lacked external validation at the polls. When Lloyd George became prime minister, also in December 1916, he was dependent not on his own party, the Liberals, but on the Conservatives, who in their turn were supportive of Robertson and Haig, at least until early 1918. The clashes between the 'frocks' and 'brasshats' were, at a generalized level, no more than remarkably clear evidence of the truisms that armies are inherently political institutions, and that their politicization arises from the very functions which they are called upon to perform.

Robertson's attitudes were shaped by two general considerations. The first was his conviction that the army must remain united: a cardinal theme for many at the time of the Curragh, the belief in its importance had only been reaffirmed by the experience of professional disunity which that incident had spawned. 'Practically anything may happen to our boasted British Constitution before the war ends', Robertson wrote to Haig on 8 March 1916, '& . . . the great asset is the army—whose value will be fixed largely to the extent by which we at the top stick together and stand firm.'[28] The second continuity was Robertson's loathing of Lloyd George, whom he memorably described—despite his own humble origins—as 'an underbred swine'.[29]

The flow of abuse went in both directions: the prime minister reciprocated by calling Robertson 'a superior clerk'.[30] But on one level the ill-feeling was surprising. At the outset of their relationship both parties were impressed by the businesslike qualities of the other. Both were committed to the maximization of the nation's resources in the pursuit of all-out victory. Lloyd George was potentially far more at odds with the ·more orthodox of his Liberal colleagues, historically committed to 'little Englandism' and wedded to a strategy of naval supremacy and fiscal rectitude rather than to conscription and a continental commitment. The tensions between Lloyd George and his generals enjoyed a high profile because they became personalized, but were ultimately of less strategic significance than might have been the case because in reality much more united than divided them.

The animosity between Lloyd George and Robertson first manifested itself in the autumn of 1916. Lloyd George had become secretary of state for war in June 1916, but found that his position had been marginalized by the enhancement of Robertson's position at the expense of Kitchener's. Lloyd George chafed at this powerlessness. But what really soured his relations with the generals was the Nivelle affair in the new year of 1917.

Lloyd George, now prime minister, agreed that the British Expeditionary Force should be subordinated to the new French commander-in-chief, Robert Nivelle. Both the British army and Britain's strategy on the western front were therefore made tools of the French government. Lloyd George had challenged the British army's professional pride and its integrity as an institution. But it was not only the generals who felt betrayed. Lloyd George had exceeded the brief given him by the war cabinet, and he had outraged George V because of the implications of what he had done for the royal prerogative. His intrigue backfired when Nivelle's offensive failed in April, and Nivelle himself was removed. Thus the ultimate result of the prime minister's effort to reduce the army's leverage was to enhance it, and the successes of the limited British offensives at Arras in April and Messines in June consolidated the soldiers' political advantage.

But the army did not use the moment either to imitate the earlier example of Sir John French or to make and unmake governments as did the German high command in the summer of 1917. The significant difference from 1915 was that the generals' political allies, the Conservatives, did not condone intervention. In May 1915 front-bench support for change was widespread; in the summer of 1917 Bonar Law, the Conservative leader, provided no such encouragement. Secondly, Lloyd George was able to lessen his dependence on the strategic advice of the army's general staff because he now had alternative institutions to which he could turn. The war cabinet may not have been an ideal institution for the direction of the war, but it provided a central focus and it constituted a final authority that was unequivocal. Hankey and his secretariat worked hard to hold it together, and to prevent its manifold subcommittees from splitting off in too many divergent directions.

Personal vendettas were therefore offset, as they had not been in 1915, by institutional strengths. Furthermore, the strategic congruence between Lloyd George and the generals operated to greater effect in 1917 than the misleading labels 'easterners' and 'westerners' suggest. Lloyd George's very advocacy of Nivelle and his offensive was evidence of his willingness to support a concentration on the western front if that could be made to constitute the most direct route to victory: Trevor Wilson has suggested that Lloyd George may even have been supportive of Haig's plan for an offensive at Ypres in the second half of 1917.[31] Equally Robertson, although he reckoned that Britain's main military effort should be on the western front, was not blind to the relationship between that front and other fronts or to the possibility of British operations elsewhere, especially in Palestine.

However, Robertson's breadth of strategic vision was to divide him from Haig, and so render forfeit the unity of the army which Robertson had seen it as so essential to maintain. Haig was commander-in-chief in France: his attention was quite properly focused exclusively on his own theatre. Robertson's task, on the other hand, was to service all the theatres in which the British army was engaged. Haig believed that it was possible to achieve a breakthrough on the western front; Robertson argued, with increasing conviction, that the battle was one of resources, that only limited objectives could be gained, and that strategy should reflect these operational constraints. The corollary of Haig's view was his belief that victory was possible in 1917; the corollary of Robertson's was that he did not expect victory until 1919.

When confronting Lloyd George, Robertson covered over these internal divisions and presented a common military front. But Lloyd George himself disliked the talk of attrition and of a long war sufficiently to be awake to Haig's promises of breakthrough and quick results. Moreover, his political antennae made him aware of the potential for division within the army, and brought him to recognize the possibility of its exploitation for his own advantage. Lord French (as he now was) still chafed at his supersession by Haig. Henry Wilson had never forgiven Robertson for refusing to pay for items allegedly left by Wilson in the commandant's house at the Staff College in 1910.[32] Neither was trusted by Haig or Robertson, but at the same time Robertson himself was increasingly isolated: the caution and limitations inherent in the realism of his strategy for 1917 began to give the impression of intellectual bankruptcy.

Henry Wilson returned to London in June 1917. Closely identified with Nivelle through his job as British liaison officer with the French, Wilson's career was threatened by the French commander's failure. He responded by warning that if he was left on the shelf ' he would get into mischief'. Unionists courted him, offering a safe seat in the Commons from which he could advocate the extension of conscription to Ireland. Some of them anticipated the collapse of the Lloyd George coalition, and hinted at even more important roles for Wilson in the government that would follow. But Wilson preferred another route to power. He was a long-standing friend of Foch, the French general whose commitment to the pursuit of victory and refusal to admit defeat bordered on the irrational.

This was the language that appealed to Lloyd George—both because it suited his temperament and because victory would be the best validation of his premiership. In party political terms, Wilson's loyalties were very different from Lloyd George's, but so too were those of most of the prime mininster's cabinet colleagues, and Lloyd George may have felt that Wilson's political credentials could buttress his position with the Conservatives on whom he depended. Strategically, the prime minister was susceptible to Wilson for the same reasons that had made him susceptible to Nivelle. Wilson may not have promised an immediate breakthrough on the western front but he did talk of stunning victories—whether he was describing knock-out blows against Turkey in 1917–18, or planning massed tank attacks against Germany for 1919. On 23 August 1917 Wilson met Lloyd George at the home of the press baron Lord Riddell, and floated the idea of an inter-allied war council, an institution whose creation would erode the powers of Robertson. In October Lloyd George brought in both French and Wilson as advisers to the war cabinet. In the circumstances of total war, strategy shaped politics; Lloyd George had split the army.[33]

By late 1917 the prime minister's own position was much more secure. The third battle of Ypres had turned both the Conservatives and the press against the generals. The staunch support of Lord Derby, the secretary of state for war from December 1916, for Haig and Robertson was offset by that of Lord Milner for Lloyd George. Milner, a member of the war cabinet, became secretary of war himself in April 1918. In November 1917 Lloyd George managed what he had failed to achieve at the beginning of the year—the subordination of the British army to a higher military authority. The Supreme War Council was created and charged with the co-ordination of the Entente's military policy. Although it lacked teeth or troops, it provided an allied structure—rather than just a British framework—to buttress Lloyd George's position. And then, towards the close of the year, on 9 December, Allenby's forces entered Jerusalem. Victory in Palestine cast doubt on Haig's and Robertson's advocacy of the primacy of the western front.

In February 1918 Lloyd George pursued his policy of dividing and ruling the military to its conclusion. He manœuvred Robertson into choosing between a reduction in his powers as chief of the imperial general staff or going to Versailles as Britain's permanent military representative on the Supreme War Council. Robertson refused both options and resigned.

Here potentially was the second major flash-point in civil–military relations in the First World War. H. A. Gwynne declared that the choice before the nation was government under Lloyd George or government under Robertson. His own preference was for the latter. He lauded Robertson as 'a son of the people, a great military genius, and an ideal leader of the nation in the great crises of this war'.[34]

But the issue was never forced. The timing of the show-down between Lloyd George and the generals was set by Lloyd George rather than by the generals. When it came, the latter had no obvious political allies. With the loss of the Conservatives, their natural backer—by dint of their shared distaste for Lloyd George— was Asquith. But Asquith, although no longer prime minister, was still the leader of the Liberal party. He could therefore hardly champion the cause of military independence from political control. Confronted with stark alternatives, Asquith—like most other British politicians—cleaved to constitutional norms, preferring civil government to military. Moreover, he wanted to see himself, not Robertson, installed as prime minister. Thus his support of Robertson against Lloyd George was muted almost to the point of silence.

The position of the generals was no stronger. There is no evidence to suggest that Robertson was anxious to push the issue to a constitutional or political crisis. He could not in all seriousness have orchestrated a coup in favour of Asquith: the latter was not a credible alternative to Lloyd George as a wartime prime minister. And his own position suffered from the fact that he was not supported by Haig. Robertson's fall was effectively the trade-off for Haig's survival. Thus Lloyd George was able to exploit the divisions within the higher ranks of the army, as his predecessors in office had played off Wolseley against Roberts. Lloyd George then solidified these divisions by appointing Henry Wilson as chief of the imperial general staff. Haig may have become doubtful of Robertson, but never had he trusted Wilson.

This basic inability of the army to master the centre stage in politics became evident again on 7 May 1918, when a clutch of newspapers—*The Times*, the *Morning Post*, the *Daily News*, and the *Daily Chronicle*—published a letter from a serving officer, Major-General Sir Frederick Maurice. Maurice accused the government of misleading the House of Commons. Two issues irked him. One was the suggestion made in a parliamentary reply by Bonar Law that the British high command had supported the extension of the British share of the line on the western front effected in January 1918. The second proved both more complex and more contentious. Lloyd George had claimed, also in a statement to the House of Commons, that the army in France was 'considerably stronger' on 1 January 1918 than it had been on 1 January 1917. In making this assertion, Lloyd George was using figures supplied to him by the War Office, but failed to note that the additional strength was accounted for not by combatants but by supporting arms. Furthermore, the choice of dates obscured the additions and subtractions which had occurred during the course of 1917 itself, and also evaded the question of the actual rifle strength on the date of the German offensive on 21 March 1918. For Maurice this was fundamental: at issue was not just the matter of accurate reporting but also the responsibility for the greatest defeat suffered by the British army on the western front since the opening German advance of 1914.

In breaching discipline by writing to the press, Maurice claimed that his obligations as a citizen had had to override his duty as a soldier.[35] But this was true only in a narrowly legalistic sense. Maurice was a highly intelligent professional soldier, young for his rank and the son of Wolseley's protégé. Robertson had brought him into the War Office as director of military operations. He had just quitted that job on Henry Wilson's suggestion, in anticipation of taking up a divisional command at the front. Three factors motivated him in May—all of them products of his professionalism. First, he was out of sympathy with the government's policy of extending the British share of the front, while diverting troops to Palestine: in this he reflected the thinking of Robertson. Secondly, he objected to Lloyd George's efforts to shuffle off responsibility for the defeats of March 1918, when Maurice's department had provided consistent warning of an impending German offensive. Thirdly, and most immediately, he felt that Lloyd George's claims were undermining the morale of the army. The implication of the prime minister's statement to the House of Commons on comparative strengths in 1917 and 1918 was that a lesser number of Germans had beaten a greater number of Britons. Thus what finally drove Maurice into the public arena was the self-respect of the British army, not the issue of parliamentary accountability.

Maurice's professionalism made him willing to engender a political crisis. But it did not lead him into the orchestration of a conspiracy. He had consulted Robertson, but to suggest that the two of them were plotting the downfall of Lloyd George suggests a degree of drive and co-ordination that is belied both by their correspondence and by their actions. Robertson backed Maurice's proposal that he should relay his unhappiness over the government's statements to his new superior, Henry Wilson. However, Maurice did not apprise Wilson of his intention to publish and Wilson did not deign to reply to Maurice's letter. Robertson also suggested Maurice should inform Asquith. But then the ousted chief of the imperial general staff changed his mind: 'not sure I like the Asquith idea after all. The case is rather one for your own judgment. The thing to do is to decide irrespective of what the results may be, unless you feel sure that the results will certainly be nil.'[36]

The tone was one of high principle, not political intrigue. Maurice's public assertions, that his letter was 'not the result of a military conspiracy', and that 'the last thing' he desired was 'to see our government in the hands of soldiers', seem entirely vindicated by the evidence. In the event, he did not consult Asquith. The former prime minister received his first warning of what was afoot from the editor of the *Morning Post*, H. A. Gwynne, on the day prior to publication. It was Gwynne, not Maurice, who hoped that the outcome of Maurice's letter would be the overthrow of Lloyd George in favour of Asquith.[37]

What Maurice looked for was some form of public inquiry into his allegations, possibly in the form of a select committee. But the war cabinet rejected this approach, preferring instead to have the issue debated as a vote of censure on the country's war leadership. Thus it was not the government's strategy but its political credibility that Maurice's letter put at stake. Similarly the House of Commons was confronted not with an opportunity to review its own powers in relation to the government through the issue of accountability but with the choice of trading Lloyd George for Asquith.

This was not at all what Maurice had had in mind. But he did nothing to try to manage events, or to steer them in a more favourable direction. Instead he scrupulously maintained his distance, absenting himself from London on the day of the debate, 9 May. Asquith gave a poor speech, Lloyd George—although somewhat free in his handling of facts—a better one. The government won by 293 votes to 106. The Liberals divided, 98 for Asquith and 72 for Lloyd George. The split had major long-term implications for the Liberal party, but more significant for Maurice was the behaviour of the Conservatives. Their backbench body—the Unionist war committee—had met on 8 May to consider its position: although sympathetic to Maurice, its members could not pursue a policy which would elevate Asquith over Lloyd George, and which would precipitate a general election on the basis of an outdated register. Thus the Maurice letter consolidated the position of Lloyd George and gave fresh legitimacy to his government.[38]

By May 1918 the army was so weakened in political terms that its only potential allies, the Asquithian Liberals, were those that it had effectively repudiated in May 1915. Part of Maurice's failure to assert himself was due to the fact that neither he nor Robertson could regard Asquith as a credible head of a wartime government: in this they did no more than reflect majority opinion. They could not attempt to direct matters because in political (as opposed to strategic) terms they had no idea of the goal towards which they should aim.

But the affair does more than demonstrate Lloyd George's ultimate success in establishing his political primacy over the military. It also shows that the urge to intervene prompted by military professionalism was contained within reasonably tight limits. Having been rebuffed Maurice did not push his case—or at least not while the war continued. He accepted the verdict of the Army Council that he be placed on half-pay with immediate effect as no less than he expected or deserved. Robertson similarly remained silent. Elsewhere in the army the ripples were barely perceptible. Henry Wilson never acknowledged receipt of Maurice's prior statement of his concerns, and disputed the veracity of his assertions.[39] In France,

Douglas Haig was equally dismissive: 'Reuter states Gen. Maurice has written to the papers. This is a grave mistake. No one can be both a soldier and a politician at the same time. We soldiers have to do our duty and keep silent, trusting to Ministers to protect us.'[40]

Ironically, some have argued that not the ministers, but Maurice, had protected Haig—that, whatever else he had failed to achieve, he had given Lloyd George a warning sufficiently salutary to head him off from any further attempt to remove the commander-in-chief.

The clashes between the 'frocks' and the 'brass-hats' in the First World War were not generated primarily by the issue of strategic choice. The relative balance between the western front and other fronts, or between the needs of the army and those of other essential wartime services, might provide the substance of the debate. But the friction which was generated acquired deeper political resonance because of the uncertainties surrounding the overall control of the government in war. The high command in general, and Robertson in particular, were doing little more than following the dictates of professional necessity as they interpreted them. The consequences were so striking because the country had entered the war without a fully developed system for its higher direction, and because between 1915 and late 1917 the government was struggling for legitimacy.

A situation that was for these reasons inherently volatile acquired explosive qualities largely thanks to the behaviour of Lloyd George. By the standards of Roberts and Wolseley, or even of Henry Wilson before 1914, the generals of the First World War were not particularly adept at, or interested in, political manipulation: the general suspicion of Henry Wilson entertained by his colleagues is evidence of that. But Lloyd George's attitudes were shaped by the previous generation. He saw the army as made up of intriguers and in-fighters: that was not just a consequence of the pot calling the kettle black, it was also a fair reflection of the army's behaviour towards Liberal governments up to and including the Curragh. The result was a prime-ministerial contempt for the generals that stood in the way of a harmonious relationship and blocked the establishment of an effective machinery for the co-ordination of strategy and policy. In the hands of Hankey, the war cabinet pointed forwards to the notion of integrated control. In the hands of Lloyd George, government fell back on the late nineteenth-century techniques of divide and rule. Lloyd George was reluctant to engage in any restructuring which might strengthen the army in its ability to tender clear and united opinions on strategic issues—a reluctance which continued until his fall from power in 1922.

The British army in the First World War was battling for the ability to shape strategy or, at a minimum, to give a coherent body of strategic advice which the government might then accept or reject in the light of its political priorities. The nature of the war—the fact that waging it permeated every facet of national life and in the end subordinated, at least for its duration, the principles of liberalism and democracy to its demands—rendered uncertain the precise boundaries of what constituted proper military concerns. But the army never saw it as its task to formulate policy on food supply, on war industries, or on the management of labour. Germany's army did. The comparison is instructive because, whatever the deficiencies of the machinery for the co-ordination of military and political advice, the structures in Britain were more sophisticated and better adapted to the needs of twentieth-century warfare than those of Germany. The latter had inherited from Prussia a model where all pivoted on the strength of the kaiser, and where the army could expand its spheres of influence precisely because of the confusion at the centre.[41] This was an example that British generals, including Maurice and Haig, specifically rejected. 'I only hope that now we have won', the latter told a group of war correspondents in Cologne on 16 December 1918, 'we shall not lose our heads, as the Germans did after 1870. It has brought them to this.'[42]

7

THE POLITICS OF GRAND STRATEGY
1918–1945: ROBERTSON'S LEGACY

The army's fundamental aim, as developed both during and after the First World War, was the effective co-operation and co-ordination of the civil and military branches of government. This was confirmed by the post-war publications of two of its most obviously 'political' generals in the war, Sir William Robertson and Sir Frederick Maurice.

Robertson's memoirs, *Soldiers and Statesmen 1914–1918*, which appeared in two volumes in 1926, have suffered in reputation not simply because any autobiographical statement is *parti pris*. They have also been eclipsed by the denigration of their author in the subsequent and lengthy farrago manufactured by his arch-rival Lloyd George, and confirmed by the phrase-turning and simplistic insights of Basil Liddell Hart. But at the time Robertson's book was commended for its objectivity and freedom from rancour. As distance diminishes subjectivity in analysis of the First World War, and as archives replace memoirs as source materials, this judgement emerges with fresh endorsement.[1]

Robertson devoted three chapters to the period before he became chief of the imperial general staff. His reason for doing this, he stated in the opening paragraph of his preface, was that before the war 'no one had sufficiently thought out . . . the organization of Government and of a High Command for war purposes'. Lloyd George's memoirs, and much of the subsequent literature, have laid stress on the civil–military tensions of 1915 to 1918. But Robertson went on to observe, with a greater perceptiveness than that displayed by many of his subsequent critics, that 'it was . . . during these early days of hurry and stress that the defects in our war machinery were the most acutely felt'.[2]

Robertson's focus was not on personalities but on systems. What was lacking in 1914–15 and what he was determined to achieve when he became chief of the imperial general staff was the creation of 'one authoritative channel', through which 'all opinions, from whatever source they might emanate' should be 'tested and criticized in relation to other plans and proposals'. This channel was of course the general staff which, Robertson argued, should be at once both the sole source of military advice to the government and 'the supreme staff authority' for the armies in the field. Thus advisory and executive functions would be united, and procedure at the War Office be brought 'into conformity with that at the front'.[3]

Robertson's aims were the entirely proper ones, in a professional context, of 'method and efficiency'. But it would be useless to pretend that these were without political implications. His interpretation of his duties was designed to prevent ministers pursuing hare-brained schemes, but it also had the practical effect of minimizing ministerial freedom of choice. The latter would be limited, in Robertson's scheme of things, to the appointment and dismissal of the service chiefs.

Robertson was not insensitive to the constitutional implications of a properly developed general staff, and he was therefore at pains elsewhere in his analysis to stress how important it was that soldiers be excluded from any wider political functions. He strenuously denied any suggestion that the war cabinet should be composed of anybody other than ministers. Military and naval chiefs should act 'as advisers and executive agents and never as members' of the war cabinet. The secretary of state for war should be a politician and not a professional soldier: Kitchener's appointment was attended 'with certain disadvantages which it is desirable to avoid'. 'It is not technical knowledge that a War Minister requires for the discharge of his responsibilities', Robertson went on, 'but the ability to obtain this information from his military colleagues, and to apply it appropriately when dealing with questions of policy and administration which call for his direction and decision.'[4]

The final pages of Robertson's two volumes were devoted to a plea for the co-ordination of strategy and policy. The ultimate responsibility for uniting the two was the prime minister's. Under him was the cabinet or 'such portion of it as may be determined': 'the activities of war embrace every element of national life, and upon the Cabinet devolves the responsibility for combining the whole military, naval, diplomatic, financial, and economic forces of the nation for the defeat of the enemy.' The professional military adviser to the cabinet must realize that,

owing to the extensive ramifications of modern war into the life of the nation, the days are gone for ever when, on the outbreak of hostilities, Ministers handed over almost entire control of the operations to military authorities and afterwards withdrew from the stage until the time arrived to negotiate terms of peace. The real head-quarters of armies in these days are to be found not in the field abroad, but at the seat of Government at home, and plans of campaign are, and must be, now analysed and criticized by civilian Ministers at the Council table in a way quite unknown a few decades ago.

The implications for both sides were fraught. Soldiers must adjust their thinking: 'without in any way becoming what are termed political Generals', they should know 'something about politics and try to understand the way in which Ministers look, and must necessarily look, at political things'. But equally ministers 'should concede to military requirements the full consideration they merit, remembering that bad strategy can never be good policy'.[5]

In his preface, Robertson was fulsome in his praise of Frederick Maurice. The admiration was mutual, and it was appropriate that these themes—of co-ordination and co-operation—should find much fuller development in a book published by Maurice in the very same year as Robertson's memoirs. *Governments and War* was not in any direct sense an attempt by Maurice to revisit the controversies of May 1918. He had already done this in *The Intrigues of War*, a series of six articles commissioned by J. A. Spender for the *Westminster Gazette* in 1922. *The Intrigues of War* was not without political significance as it was a sustained critique of Lloyd George's management of strategy published when Lloyd George was still prime minister. But Maurice's opening paragraph made clear the distinction between his own personal differences with Lloyd George and the more general question of the army's relations with politicians:

'It is customary for soldiers to damn politicians. While this is very natural, it has never seemed to me to be a very useful practice, for no soldier has yet invented an efficient substitute for the politician. It is more profitable, therefore, to try to make politicians better than to condemn them to perdition.'[6]

This more positive task was that embraced by *Governments and War*. Developed from his Lees Knowles lectures of 1925, the book took as its principal case-study not the First World War but the American Civil War. Maurice specifically rejected G. F. R. Henderson's attack on political intervention in the conduct of that war which had been endorsed with so much enthusiasm by Wolseley.[7] He disliked the soldier's tendency to see the 'damned politician' as a nuisance, and emphasized that the statesman 'is, even from the purely military point of view, the indispensable ally of the soldier, and should be treated as such'.[8]

His own perspective was Clausewitzian, albeit in reworked form. He likened the relationship between politics and strategy to that between a parent and child.

It is the duty of policy to choose the road for strategy, to set it on its way, to provide means sufficient for the journey, to give timely counsel, to watch the youngster's progress carefully, to be prompt to give a hand should he stumble, to be ready to turn him in a new direction should a change of course seem necessary or opportune, but to resist the temptation to interfere save as a measure of real emergency, and then to make interference as little obvious as may be. It is no easy task to be a wise parent.[9]

Maurice's theoretical formulation, like Clausewitz's, implied the subordination of war to politics. But practice for him, as also for Clausewitz, was more a matter of fusion. 'Modern war', he stated, 'demands not Napoleon's man but a partnership between the statesman and his military commander. I use the term military in its widest sense to include all armed forces. The statesman must be the senior partner, and if the partnership is to be effective its members must have confidence each in the other, must be sufficiently acquainted with the whole business to understand the needs and difficulties of each, and the senior partner must know when and how to leave his fellows to their tasks, how to direct without interference.' However, the obligations created by integrated command were mutual. 'If the junior partners are to do their business they must know on what principles and by what methods their senior will act.' They had to rid themselves of the misconception that the direction of armies and navies was a matter best left to soldiers and sailors. 'To-day at least we should be aware that [the conduct of war] means the direction for a special purpose of the whole power and resources of the nation. This is clearly not a matter to be left to soldiers or sailors, nor would any responsible soldier or sailor desire it to be so left.'[10]

The requirement that each side understand the other carried an implicit assumption of greater equality between the parties than was normal in the relationship between ministers and civil servants in other government departments. The statesman had to realize that in war the professional military adviser was not a technocrat comparable with the civil servant. First, a general could not be anonymous: 'he is a public figure, and the public regard him as directly responsible to it for the employment of the forces in which serve the sons, brothers, sweethearts, and husbands of the nation.'[11] Secondly, the soldier could not carry out a plan with which he was in disagreement. 'In war the moral element is predominant. It is not in human nature for a man to prosecute wholeheartedly a plan, which involves great risks and the lives of others if he does not believe in it absolutely.'[12] Thus the politician had to support the general, and had to make him aware of a continuing confidence in his abilities. This was Lincoln's service to Grant. Lincoln was able to do this because 'he thought of nothing but what was the right military policy, when he had every temptation to urge what was politically expedient'.[13] If the head of state found himself unable to sustain his commander-in-chief in comparable fashion, then the only sanction open was to dismiss him.

For Maurice, as for Robertson, this relationship could only function without friction if it rested on a system whose rules were fully understood and subscribed to by all parties. Clear instructions from the statesman to the soldier were essential, but the instructions should be general rather than specific.[14] In order to ensure consistency between means and ends, and between the operations in one theatre and another, the advice on which this policy rested should be derived from a single source, not from several. In *The Intrigues of*

War, Maurice had attributed 'most of our troubles in the present and other wars' to the fact that 'British ministers have never been able to grasp the difference between the opinion of individual soldiers, however distinguished and experienced, speaking for themselves alone and the opinion of a responsible military adviser speaking as the head of an organization established and designed for the preparation of considered and co-ordinated views on military questions'.[15] Maurice's Robertsonian credentials were therefore manifest in his support of the general staff system established in December 1915. 'What Sir William Robertson's proposals did . . . was to define clearly the respective functions of Ministers and soldiers in war—to set forth the terms of the partnership.'[16]

In November 1924, a few months before Maurice addressed his Cambridge audience, a subcommittee of the Committee of Imperial Defence, chaired by Lord Salisbury, had reported on the organization of imperial defence. It proposed that a committee of the professional heads of the three services (the RAF having been added in April 1918) should be established on a permanent basis. Such a body had already begun to function in an *ad hoc* way in 1923, significantly after Lloyd George's departure from power. Each of the three chiefs of staff would continue to have an individual responsibility for his own service, and indeed it was in this capacity that his executive authority resided. But in addition they would have the collective task of advising on defence policy as a whole, 'the three constituting, as it were, a Super-Chief of a War Staff in Commission'.[17] In 1926 the chiefs of staff committee was reinforced by the addition of a joint planning committee, made up of the three services' directors of plans. The chiefs of staff were serviced by Hankey's secretariat, which also serviced the revived Committee of Imperial Defence.

When the war had ended, many in Britain had called for the establishment of a ministry of defence. The creation of a third service reinforced the view that only thus could effective co-ordination be achieved. Its advocates included Winston Churchill, who in 1919 became secretary of state for both war and air, and Rear-Admiral Sir Murray Sueter, the leading naval exponent of air power who in 1922 introduced a bill for the creation of a ministry of defence. A principal opponent of such schemes was Hankey. His fear—and indeed that too of Lloyd George while he was still prime minister—was that a minister of defence would usurp the powers of the prime minister.

An indication of how far integration had already proceeded was the comparative irrelevance of constitutional questions to this debate. Although present, they were not decisive. Hankey's protection of the powers of the prime minister, together with his development of the committee system and its accompanying secretariat, could be seen as part of a process elevating the prime minister to presidential status. Equally sinister to constitutionalists was the role of military men in that machinery. The chiefs of staff committee did not even balance servicemen with civilians, and the creation of such a weighty professional body threatened to ensure military domination in practice, if not in theory. Winston Churchill's reassurances to the House of Commons in November 1919 smacked of insouciance to those of liberal dispositions: 'The initiative in Service matters must in the main come, and as a general rule comes, from the professional head. He plans, he outlines, he proposes. The Minister examines, criticises, suggests, discusses with his Board or Council, *and finally approves*. That is the right way.'[18]

Increasingly, therefore, the issues which preoccupied would-be organizers of defence were no longer the constitutional safeguards against the military, but the requirements of strategic direction in war. So effectively had the dividing line between the proper spheres of military and political responsibility been eroded that the important debate was over matters of practicality. Specifically, the need was to establish a machinery for the higher direction of defence in peace that would be appropriate to the needs of war. The Salisbury committee had recommended the appointment of a cabinet minister to deputize for the prime minister as chairman both of the Committee of Imperial Defence and of the chiefs of staff. The attraction of this scheme was that it recognized that, while in peacetime defence would not be a first call on the prime minister's time, it would be in the event of war. Its critics saw such an arrangement as a botched job—a half-way house to a full ministry of defence.

Maurice's contribution to this debate reflected a distinction also made by Robertson. Britain needed two systems, one for peace, when the normal functions and priorities of a democracy would dominate, and another for war. In the first instance, the prime minister's concern could not be with matters of strategy, and he could therefore leave the chairmanship of the defence committees to a deputy. It might even be permissible to create a ministry of defence. But in war the prime minister must lead. The direction of strategy would be his main preoccupation, just as it would be the task to which the entire nation's effort would be bent. Maurice therefore argued that the chiefs of staff should become directly answerable to the prime minister in time of war. In 1927 an Imperial Defence College was established with the intention of fostering a tri-service view of defence: between 1929 and 1933 this became the forum in which Maurice expanded and expounded the views on defence organization which he had first adumbrated in *Governments and War*.[19] Through Maurice's teaching, the inter-war orthodoxy on civil–military relations, at least in service circles, was Robertsonian.

Maurice's scheme was effectively that implemented in the Second World War. In 1936 Sir Thomas Inskip was appointed minister for the co-ordination of defence, a belated concession to the Salisbury committee's recommendations on the matter of chairmanship. But Inskip was a minister without a ministry. His functions were the peacetime ones of balancing service procurement bids rather than the wartime role of strategic direction. When the war broke out, the option that Hankey had fought so zealously to preserve, a war cabinet modelled on that created by Lloyd George, was promptly set up. But its redundance, evident even before Churchill became prime minister in 1940, was confirmed thereafter. Churchill was his own minister of defence as well as prime minister. The individual service ministers were sidelined, confined to the administration of their particular armed force, and deprived of strategic responsibility. The chiefs of staff committee became the prime minister's instrument for the control and direction of the war.

For Churchill and for his subordinates the memory of civil–military friction in 1914–18 created a heavy burden of expectation. After the United States entered the war, the report that George C. Marshall, the American chief of staff, was reading Robertson's *Soldiers and Statesmen* prompted a ministerial stampede for the book.[20] Ismay, Hankey's successor, testified to his own determination to avoid a repetition of the division between 'frocks' and 'brass-hats'. Like others of his generation, he devoured Henderson's *Stonewall Jackson*. But his own Staff College training (at Quetta, rather than Camberley) reflected not the strictures of Henderson but the orthodoxy enunciated by Robertson and Maurice: 'It is essential that the statesman and the soldier should be in the closest touch with each other, and that they should work together frankly, openly and loyally . . . The soldier must realise that war is the instrument of policy, and that, so far as any subservience is necessary, he must be subservient to the statesman.' Ismay's memoirs presented a verdict entirely in accord with his prescription. 'The War Cabinet', he wrote with a blandness characteristic of the volume as a whole, 'was a band of brothers, and . . . relations between Churchill and his official advisers, both civil and military, were characterised by mutual understanding, esteem and affection.'[21]

In so far as Ismay's account is entirely honest, its portrayal of harmony was largely a reflection of his own efforts as interpreter, facilitator, and mediator. For, in reality civil–military relations in the Second World War were for much of the time as potentially fraught as those in the First. The vocabulary of togetherness suggested by Robertson and Maurice—convergence, co-ordination, fusion, integration—belies the seething tensions and simmering discontents which were no less deeply felt for not having exploded into public gaze. Churchill was even less disposed to honour the 'system' advocated by Robertson than Lloyd George. This was not just a matter of irregular hours and long nights; it was also, more seriously, an enthusiasm for alternative strategies and for taking counsel from officers outside the approved channels.

It would be tempting to argue that the civil–military battles of the Second World War never enjoyed the high public profile of those in the First because the army had now been neutralized—that Lloyd George had fought Robertson and the general staff, and had won. Such a thesis might find reinforcement in the post-1918 injunctions of Robertson and Maurice that soldiers should recognize the sovereignty of statesmen. And it would, no doubt, conclude that the integration of civilians and soldiers in the higher administration and direction of war, especially through the chiefs of staff committee, had satiated the political ambitions of the army.

However, such explanations, while containing partial truths, would be guilty, once again, of underestimating the political instincts of an army motivated by professional considerations and fired by its *amour propre*. A clear demonstration of the fact that these survived the inter-war years intact was the fate of Leslie Hore-Belisha, secretary of state for war, in January 1940.

Hore-Belisha was at once both 'insider' and 'outsider': a Cliftonian (like many generals, including Haig) but a Jew; an officer in the First World War but in the Royal Army Service Corps rather than a line regiment. Between his appointment in 1937 and the outbreak of hostilities in 1939 he set about the reform of the army with a vigour that threatened to cast Cardwell and Haldane in the shade. But, unlike them, he failed to enlist the support of the rising generation of senior officers. True, he purged the Army Council, and secured his own nominee as chief of the imperial general staff—Lord Gort. The choice, however, was a poor one. Gort had won the Victoria Cross and aspired to command the army in battle; he had little enthusiasm for the committee work engendered by the army's principal staff job. None the less he endeavoured to devote four hours a day to the reading of books on military affairs.[22] The effort was perhaps a reflection of where the army perceived the real impetus behind Hore-Belisha's reforms to lie—in the pen of the military correspondent Basil Liddell Hart.

Although Liddell Hart had backed Gort for the job which he now held, Gort himself could hardly look with favour on the disproportionate influence which Liddell Hart was believed to enjoy. This was not Wolseley or Haig: Liddell Hart had been invalided out of the army as a captain, after less than two years' active service at the front. But he never allowed this comparative lack of experience to temper his views with a becoming modesty. Expressed with fluency, and clothed with metaphors which aided comprehension by the amateur but obscured the many-sidedness of the problems that confronted the professionals, Liddell Hart's writings commended themselves more to those outside the army than to many of those within it. His opinion of the British high command had been irretrievably coloured by Charles Callwell's life of Henry Wilson, published in 1927. Liddell Hart associated all British generals with its revelations of political intrigue and personal ambition. William

Robertson in particular became the butt of his explanations for the failings of British strategy in the First World War. He saw the general staff as having committed Britain to war on the continent of Europe and to a Continental style of war. Possessed of a romanticized view of the great commander, whose attributes spanned time, he saw generalship as a matter of moral forces, not material, unchanged by industrialization, new technology, and conscription. Staffs and staff work were not part of his mental picture of war. For Liddell Hart, Robertson's 'system' had diverted British strategy into the wrong channels, and had killed imagination and enterprise.

The casualties of Hore-Belisha's purge in December 1937 included the chief of the imperial general staff (Sir Cyril Deverell), the adjutant-general, and the director of military intelligence. The clear-out could be interpreted as the removal of the old guard, wedded either to the defence of India or to the methods of 1914–18, and rendered administratively incompetent by advancing years. But the real problem was not senility but strategy. Neither Hore-Belisha nor Liddell Hart supported the creation of a field force for Continental war; the general staff as a unit did. Thus the successors of those who had been removed, though chosen by Hore-Belisha himself, proffered advice that was little more congenial than that tendered by Deverell. Anxious to support the Chamberlain government's priorities, which put the navy and the RAF ahead of the army, Hore-Belisha invited Lords Weir and Trenchard to meetings of the Army Council. Weir was the government's industrial adviser and Trenchard had been chief of the air staff; Hore-Belisha even considered making him chief of the imperial general staff.[23] Consulting outside experts like Liddell Hart, taking soundings from senior servicemen in defiance of the conventional channels—these were methods to clip the powers of the general staff.

Henry Pownall, the director of military operations and the most outspoken Robertsonian in the War Office in the late 1930s, concluded that Hore-Belisha's aim was 'to bitch up the higher organization of the Army in order to make the position of CIGS impossible'.[24] By June 1939, with war looming, relations between Hore-Belisha and Gort had become so bad that the former had only sent for the latter once in the previous month,[25] and the latter preferred to communicate with the former indirectly, via his deputy, Sir Ronald Adam.

One of the officers drawn into this process of dividing and ruling was Sir Edmund Ironside. Himself passed over for the post of chief of the imperial general staff, he was now inspector-general of overseas forces and therefore the putative commander-in-chief of the British Expeditionary Force. But on 3 September 1939, the day of Britain's entry to the war, Hore-Belisha appointed Ironside chief of the imperial general staff and gave Gort the command of the British Expeditionary Force. Gort was delighted by the swap, and took Pownall with him to be his chief of staff. Any continuity between pre-war planning and the direction of strategy on mobilization was thus immediately forfeit.

System was secondary. Civil–military relations in 1939 were being described by many informed observers in terms evocative of the First World War. Sir James Grigg, who took over as permanent under-secretary at the War Office that summer, was aghast at the divisions he encountered: 'On a minor scale', he wrote to his father on 25 May, 'H-B feels very much like L.G. about the soldiers, and the soldiers certainly feel about H-B, what their predecessors in 1918 did about L.G.'[26] Pownall used the same example: in June he concluded that Gort's best line was to behave as Haig had done when hounded by Lloyd George—to call the ministerial bluff by refusing to resign, so daring the government to risk its credibility by dismissing him.[27]

As Gort and the British Expeditionary Force took up positions in France already made familiar in battles just over twenty years before, the temptation to revisit the past was inexorable. Facing Belgium once again, the French and British planned to advance rapidly into the Low Countries so as to seek an encounter battle there, rather than fight defensively on the territory of northern France.

Hore-Belisha visited the army in late September and again in November. The creation of the war cabinet had left him, as it had left his predecesssors in 1916–18, unsure of his role: in November he complained that the service ministers' contribution was marginalized by the importance in strategic direction of the chiefs of staff.[28] He compensated by making recommendations in operational matters. Pownall was aghast because Ironside appeared to collude in this civilian intrusion into purely military matters. Above all, Hore-Belisha became fixated with pill-boxes—with the ease of their construction, with the idea that the army had not got enough (although it planned to fight an encounter battle), and with the notion that it was dragging its feet in building them. After his second visit to France, he told the war cabinet—when Ironside had left the room—that during his tour he had seen only two new pill-boxes being constructed. On 27 November word reached Gort's headquarters that the prime minister was 'very perturbed at the reported weakness of the British sector of the line'.[29]

Before Hore-Belisha's second visit to France, the army held him guilty on two charges—the breach of Robertson's system, and civilian intrusion into purely military affairs. Now a third was added. The secretary of state's comments in London were undermining morale at the front. Pownall looked to Ironside to back his field commanders, but found him wanting: the chief of the imperial general staff could, Pownall significantly concluded, have reread Robertson's *Soldiers and Statesmen* with profit.[30] But Pownall underestimated Ironside's own mounting frustration with Hore-Belisha. At the end of November, Ironside crossed the channel to see matters for himself. He was outraged by what he found: 'One and all were cursing Belisha for accusing them of not having done any work.'[31]

The refrain which greeted Ironside at Gort's headquarters was orchestrated by Pownall. Awaiting his arrival, Pownall wrote: 'We are now here all facing *West*, to meet the more dangerous enemy there.' Pownall's plan was to win over Ironside so that the chief of the imperial general staff and the commander-in-chief could combine to oust Hore-Belisha: 'I pray heaven this may be the last of him. I am doing my best to make it so.'[32]

Ironside himself denied that there was a military 'cabal' against Hore-Belisha; Sir John Kennedy, then deputy director of military operations, concluded that there was.[33] Certainly Pownall used the familiar tools of military interventionism. He was one of two generals identified by Liddell Hart as having enlisted the *Sunday Express* in support of the army's case.[34] More importantly Pownall deemed it vital that the king should know what was going on. He found time between Hore-Belisha's and Ironside's visits to get back to London in order to give George VI's private secretary, Sir Alexander Hardinge, a full and frank report on the soldiers' view of the secretary of state.[35] Hardinge was an old friend of Gort and, like him, had served in the Grenadier Guards. Thus alerted, the king summoned Ironside to Buckingham Palace on his return from France, and a couple of days later himself joined the procession of visitors to Gort's headquarters. Pownall was satisfied with the outcome: 'Both the King and Hardinge are under no illusions about Hore-Belisha and realize that he must go. We did not fail to keep them fully informed of all the details of H-B's recent disgraceful behaviour, and there's no doubt we have the Palace on our side against him. The King, when I sat next to him at dinner, went so far as to ask me who, in my view, should replace H-B at the War Office.'[36]

George VI was convinced of the 'angry discontent seething in the B.E.F.', and spoke to the prime minister about it.[37] Now it was the turn of Chamberlain to add to Gort's flow of visitors. He told Pownall that he was anxious to reassure the commander-in-chief that the war cabinet had full confidence in him; Pownall was quick to respond that '*our* trouble was that we did not have confidence in getting support from H-B'.[38]

Chamberlain decided to move Hore-Belisha from the War Office to the Board of Trade. Hore-Belisha refused the post offered to him and left the government. The prime minister had not revised his favourable view of Hore-Belisha as a minister or of his achievements in office. But he was fearful that the frictions of the First World War were about to be rekindled. Pownall described Hore-Belisha after his fall as 'an enemy at our back', who had modelled himself on Lloyd George and was using 'Lloyd George tactics'.[39] These were exactly the sentiments which Chamberlain was anxious to douse. 'In wartime nothing could be worse than perpetual friction and want of confidence between the Secretary of State and the C-in-C in the field.'[40]

Thus, in their first brush with the politicians, the soldiers of the Second World War had successfully exploited the memory of the First so as to ensure freedom of movement in their own areas of responsibility. Professional preoccupations had promoted political intrigue, and the latter had been furthered by methods following familiar grooves.

The sanctity of military judgements in strategic matters was honoured not only by Chamberlain. Churchill too made obeisance to it. Confronted with a succession of defeats and challenged in parliament, he defended himself in the winter of 1941–2 with a portrayal of his relationship with the chiefs of staff that paid implicit homage to Robertson. 'I do not conduct this war from day to day myself; it is conducted from day to day, and in its formal outlook, by the Chiefs of Staff Committee . . . I do not think there has ever been a system in which the professional heads of the fighting services have had a freer hand or a greater or more discreet influence or have received more constant and harmonious support from the Prime Minister and the Cabinet under whom they serve.'[41]

The only truth which this statement revealed was that defeat had no authors. Churchill's account of the relationship between himself and his chiefs of staff enabled him to shift responsibility for successive set-backs. It did not constitute an accurate portrayal of the formulation of strategy. The chairman of the chiefs of staff was not the country's principal strategic adviser. Churchill had appropriated that job for himself. The chairman was the convenor of a group of experts on the three individual armed forces. The post acquired wider functions with the appointment of Alan Brooke as chief of the imperial general staff at the end of 1941, but Brooke was still not the initiator of strategy.[42]

The army knew Churchill's reputation as an amateur strategist and inveterate meddler from the First World War. The failed expedition to Norway in the New Year of 1940 smacked 'all too alarmingly of Gallipoli'.[43] His interventions in the battle for France in May combined a lust for tactical involvement with an insensitivity to practical constraints familiar to those who had served under him at the Admiralty in 1914–15. The marginalization of the war cabinet and the fragmentation of the chiefs of staff into its service components meant that Churchill was 'not only advocate, but witness, prosecutor and judge'. Britain's strategy was personalized, with military opinions being distorted by the prime minister's oratory, with Churchillian directives being sent without professional advice, and with the lust for immediate action always threatening the formation of longer-term strategy.[44] Even when the roll-call of defeats began to give way to a sequence of victories, professional opinion was unrelenting. In September 1944 Alan Brooke, exhausted by the bruisings and buffetings of his relationship with Churchill, and privately feeling 'it would be a Godsend if he could disappear out of public life', wrote: 'Three quarters of the population of the world imagine that Winston Churchill is one of the strategists of history, a second Marlborough, and the other quarter have no conception of what a public menace he is, and has been throughout the war.'[45]

On 27 May 1940 Ironside was replaced as chief of the imperial general staff by Sir John Dill: a square peg was removed in favour of a round one. He had held staff appointments throughout the First World War, had been commandant of the Staff College between 1931 and 1934, and was director of military operations thereafter. In 1940 he was one of two corps commanders (Brooke, his successor as chief of the imperial general staff, was the other) implicated in the fall of Hore-Belisha. He was a Robertsonian twice over—in his immersion in the general staff system, and in his recognition that professionalism could promote interventionism. But, as with Robertson, intervention was reluctant: he had no enthusiasm for intrigue *per se*, and he preferred to convince by methodical written exposition rather than by fierce oral exchanges. His ways were not Churchill's.[46]

In the clashes these differing temperaments generated, Dill took consolation in the example which Robertson had set. Just after midnight on 4 December 1940, a slumbering John Kennedy, now director of military operations, was stirred by Dill. 'I cannot tell you how angry the Prime Minister has made me', Dill confessed. 'What he said about the Army tonight I can never forgive. He complained he could get nothing done by the Army. Then he said he wished he had Papagos [Greece's commander-in-chief] to run it.' Kennedy reminded Dill of the frustrations with which Robertson had to contend in his dealings with Lloyd George. However, there was a significant difference between the positions of Dill and Robertson. The latter was buttressed by the knowledge that at least one commander in the field, Haig, had gained sufficient prestige to support the army in its battles with politicians.[47] In late 1940 there were not even putative victories to refurbish the reputations of Britain's generals.

Within days of Dill's distress the army seemed to have found its great leader. Wavell opened his offensive against the Italians in the western Desert with stunning successes. But the army's professional and political pride was not to be sustained. First there was the question of the primacy of the Middle East as a theatre of operations. For Dill it had to rank below the defence both of the United Kingdom and of Singapore. For Churchill, the fact that Libya provided the opportunity for attack was sufficient unto itself. Differing priorities in relation to the theatres of war awakened further uncomfortable parallels with the recent past. Secondly, there was Churchill's pressure to send troops to Greece—a pressure to which Wavell and Dill both succumbed. Thus the theatre of operations in which the army had achieved its successes was marginalized by Dill himself, and the commander who had acquired that lustre was also discredited.

Privately Dill's thoughts veered from the Robertsonian model: 'One cannot condemn Henry Wilson so heavily as one used to now that one has had first-rate experience of politicians.' Publicly he still followed Robertson's precepts. He advised Churchill to back Wavell or sack him. Churchill, fearful of the political implications of having a distinguished general unemployed at home, shipped Wavell off to India.[48] Dill himself considered resignation, consulting Hankey on the precedent set in 1918 by Robertson himself.[49]

Churchill's relations with his strategic advisers changed early in 1942. In part this may have been a response to an attack on his machinery for the conduct of war, mounted from the Lords in the wake of Singapore's fall by Chatfield and Hankey. Hankey in particular was angered by Churchill's failure to use the war cabinet. But parliamentary thrusts against the prime minister lacked substance when no serious challenger to his primacy was in evidence. Much more significant over the long term was the replacement of Dill by Brooke.

Brooke managed Churchill in two ways. First, he answered back. He protected the army's field commanders, interposing himself between them and the prime minister, injecting realism into the latter's schemes, and protesting when their achievements were called into question. Secondly, he became chairman of the chiefs of staff and held that post for the rest of the war. He did not seek to develop a tri-service perspective; he argued that the strength of the chiefs of staff committee lay in the fact that its members were the executive heads of their services. But he did see the need for more effective co-ordination, and in particular favoured the appointment of single supreme commanders for each theatre, irrespective of service.

In all this he did not deny the political instincts to which his professionalism gave birth. At the Washington conference in May 1943, the prime minister's shifting strategic enthusiasms drove Brooke to distraction. At a meeting at the White House, Churchill repudiated a paper prepared by the chiefs of staff and agreed to by him. The suspicions which he engendered in American minds did, in Brooke's words, 'untold harm'.[50] In March 1944 the chiefs of staff favoured a strategy for the Pacific that would contribute to what was primarily an American effort and which would build on relations with Australia. Churchill pushed a different orientation, designed for the reconquest of Malaya and Singapore. On this occasion, the chiefs of staff contemplated resignation *en bloc*.[51]

In the political history of the army, the most intriguing question about the Second World War is therefore a counter-factual one. Robertson may have moderated and channelled the political instincts to which the army's professionalism gave rise, but, as the case of Hore-Belisha shows, those political instincts were not therefore extinguished. Churchill contravened every notion of system in civil–military relations to which Robertson had given expression. And yet the army did not move against him. Why not? Why was its frustration confined to the diaries and correspondence of men such as Brooke, Kennedy, and Pownall?

An important difference between the two world wars was that until 1944 no front in the Second World War possessed the primacy that was held by northern France in 1914–18. Thus the battle for the control of strategy did not have a consistent focus around which debate could polarize. A corollary of this was that the army was not the dominant service as it had been under Haig and Robertson. Indeed Churchill's ability to survive despite the defeats of 1941 and 1942 can be attributed to his support of strategic bombing and his sense of the centrality of the battle of the Atlantic. By ensuring that he occupied the pole position between the services, he could make inter-service rivalry work to his own advantage. The army's belief that it was being denigrated in the process[52] was of little significance when its lack of success in the field ensured that its reputation was at a discount.

The army was therefore deprived of political clout. And it recognized that, unlike Asquith in 1915, Churchill—even in 1941–2—had no serious parliamentary opposition with which to contend. Kennedy noted that a vote of censure on the conduct of war was put down for debate in the House of Commons on 25 June 1942. Churchill carried the House on 2 July by 475 votes to 25.[53] Thus there were no opposition coat-tails to which the army could cling.

Churchill's leadership therefore offset Churchill's strategy. It has been argued in the previous chapter that in 1916–18 more united Lloyd George and the generals than the rhetoric of their disputes suggested: both were committed to the all-out pursuit of victory. Exactly the same point can be made about Churchill and the generals in 1940–5. Churchill's political survival rested on his being the public embodiment of the will to win. This, for the soldiers, elevated him to the point where professional frustration gave way to political respect. Brooke's diatribe against the prime minister's strategic judgement of September 1944 went on, appropriately enough: 'Without him England was lost for a certainty.'[54]

Brooke's public silence during the war was a vital element in ensuring that the relationship between Churchill and his chiefs of staff did not rupture. The strategy of the Second World War was not directed in accordance with the precepts of Robertson. Like that of the First, it was thrashed out by way of tantrums, tensions, and compromise. It was shaped by the clash of personalities. But, vitally, it was seen in a totally different light—to be the product of system and team-work. The effect of Brooke's personal contribution to civil–military relations could therefore be institutionalized. The notion of integration survived its greatest ordeal.[55]

8

THE POLITICS OF COUNTER-INSURGENCY

Integrated control was shaped by the needs of European war. For most soldiers most of the time war on this scale was not in itself a politicizing experience. Because the needs of the service were more often sublimated than thwarted, professionalism in its broadest sense did not promote politicking. World war justified giving soldiers a comparatively free hand. Thus the central premiss of integrated control—that soldiers were the equals of politicians in practice, even if their subordinates in theory—was easy to sustain. Furthermore, the Clausewitzian approach to strategy, that war was an instrument of politics, seemed remote to all but the most senior commanders: even at the higher operational levels of corps or army command the political consequences of military action were distant. For many of the participants in the two world wars, war for much of the time was indeed an end in itself.

However, the principal factor which had made for volatility in civil–military relations in the nineteenth century—the empire—did not cease to operate in the twentieth. The Curragh was not the end of the line. Through the acquisition of the German colonies and by virtue of the mandates established under the Versailles peace settlement, the British empire stood at its greatest ever extent at the conclusion of the First World War. In policing the colonies, and above all in the techniques of counter-insurgency, the political consequences of military action were direct, even down to company and platoon level. The experience of withdrawal from the empire was as potentially politicizing for the army as had been the processes of its conquest.[1]

At one level the axioms of counter-insurgency were identical to

the precepts of major war. The fusion of civil and military authorities to create the machinery of co-ordinated control became a
cardinal precept in the former as in the latter. But in the first
case practice has had even greater difficulty in aping theory than
in the second. If the situation is interpreted not as one of war or
revolution but as one of terrorism or civil disorder, then the primacy
of civilian control can be an important part of the democratic and
liberal credentials of the government. But the assertion of civil
control necessarily vitiates the virtues of civil–military co-ordination, so jeopardizing co-operation between the army and the police.
Two practical problems follow. The first is the difficulty of collecting
and assessing intelligence if the security forces are divided, and even
in competition with each other. The second is the correct application of force. Minimum force is a key feature of British policing: it
reinforces the liberal application of the law, and removes any
justification for a terrorist backlash. Maximum force is the natural
response of the soldier: its rationale in this case is that a show of
strength implies resolution and thus constitutes its own deterrent.

In 1919 the British army confronted these problems in two places,
both of them prime forcing grounds in its nineteenth-century politicization—Ireland and India. The experiences were significant for
the future in 'the lessons' they were perceived to contain. But the
nature of those experiences was shaped by the effects of the past.

In the case of Ireland, it was the legacy of the First World War
that bedevilled the army. There were of course the problems of
adaptation: the experience of guerrilla warfare had been forfeit to
the pressures of mass industrialized war. The troops themselves
were tired or ill-trained, caught between the processes of demobil-
ization and post-war reduction on the one hand, and the expansion
of imperial policing on the other. But the most important consequence of the Great War was Lloyd George's distrust of the generals. It undermined civil–military co-ordination and hence worked
against the formation of clear policy. Ireland, the prime minister
said, 'was a policeman's job supported by the military and not *vice
versa*. So long as it becomes a military job only it will fail.'[2]

The trouble with this prescription was not just its derogation of
the soldier. It was also the ambiguous status of the Irish police. The
Royal Irish Constabulary was commanded by a gunner officer,

Major-General H. M. Tudor. The director of intelligence was another job nominally a policeman's but held by a soldier. In April 1920 Tudor expanded the RIC by recruiting former regular soldiers, the Black and Tans, and in July he added another body of wartime veterans, the Auxiliaries, also commanded by a soldier, Brigadier-General F. P. Crozier. Crozier was notorious as a 'fire-eater'. Neither Tudor nor any of his military appointees possessed backgrounds appropriate to the task in hand, and they made little effort to reorientate themselves. They saw the Irish people not as potential allies, but as their new enemies—a substitute for the Germans. Thus the RIC itself was a potential law-breaker. Ostensibly, the fusion of soldiers with policemen seemed to create the conditions for co-operation. In reality it confused the functions of the two, generating a body that was para-military but was independent of the army.

Equally symptomatic of civil–military fusions and equally disappointing in its outcome, albeit for different reasons, was the appointment in 1920 of General Sir Nevil Macready as commander-in-chief in Ireland. Macready had been adjutant-general in the Great War, but he had also been commissioner of the Metropolitan Police since 1918. He knew Ireland and he favoured home rule. He disliked Tudor, and he was appalled by the state of the Royal Irish Constabulary. The army found itself protecting the police or dealing with the consequences of its excesses. The former rather than the latter became the force for moderation. Friction, not co-operation, became the order of the day between the two services. For Macready, professionalism increasingly manifested itself in institutional self-respect: the honour of the army ranked higher than the success of British policy.

Macready favoured the introduction of martial law. The chief secretary for Ireland was doubtful, arguing that 'reprisals would not affect public opinion in England or the stability of the Government'. For Macready such considerations were minor 'in comparison with the good name of the army'. When martial law was eventually established, in December 1920, it was applied to a restricted area and not to Dublin. Moreover, although he acknowledged the theory of political supremacy, Macready was frustrated by the controls within which he was still required to operate. Told that he must seek prior sanction from the government for any steps he wished to take, he responded that 'For the soldier such a position was impossible.'[3] In February 1921, Sir Warren Fisher, the head of the civil service, made two recommendations—that Macready be appointed civil–military supremo, and that martial law be extended to all Ireland. Macready did not support the first suggestion: he saw the reform of the Royal Irish Constabulary as a major task in its own right. But he did of course favour the second. Lloyd George, on the other hand, fearful of the elevation of the military, baulked at the proposal. Britain slithered to defeat.[4]

The idea that the army enabled Britain to hold Ireland was inimical to the concept of the United Kingdom. The idea that the army enabled Britain to hold India had been fundamental to the Raj throughout the nineteenth century, and received powerful endorsement in 1857. The First World War reminded the government of India of the possibility of rebellion, fomented by Indian exiles in the United States and in Germany, and directed towards Bengal or from Afghanistan towards the north-west frontier. But at the same time the army in India was weakened by its commitments to operations outside India. Pragmatism, in the shape of the Montagu–Chelmsford reforms, therefore nudged British rule towards greater political liberalism. Two concepts of governance were potentially at odds. On the one hand was the belief that Britain was unwelcome in India and in the last resort could only hold it by force of arms. On the other was the conviction that enlightenment would reveal the universal acceptability of British values. The commander-in-chief in India, General Sir Charles Monro, said of the secretary of state for India that 'When I am with Mr Montagu I feel that I am walking with a man who is steadily edging me towards a precipice and when he gets me near enough will push me over.'[5]

On 13 April 1919 Brigadier-General R. E. H. Dyer ordered his troops to open fire on a large crowd assembled in the Jallianwala Bagh in Amritsar. In ten minutes they discharged 1,650 rounds, leaving 379 dead and over 1,200 wounded. Dyer personally directed the operation, and followed it up with measures calculated to humiliate the population, whether innocent or guilty. Dyer did not have the sanction of martial law when he acted. He did however have the authorization of the deputy commissioner of Amritsar to use force of arms 'if necessary'. In the subsequent controversy he enjoyed the full backing of Sir Michael O'Dwyer, lieutenant-governor of the Punjab.

O'Dwyer was persuaded that revolution was imminent. He had kept the Punjab in order throughout the First World War despite the threat of German-inspired insurrection, and in the process he had managed to sustain its status as a major recruiting ground for the army of India. The potential links between disorder and external threats were corroborated immediately after the massacre at Amritsar, when war broke out with Afghanistan. Dyer's values were those of Anglo-India. His father had been a brewer in Murree; his career and most of his life had been passed in the subcontinent. The issue that he felt confronted him in the Jallianwala Bagh was therefore simple—that of war or peace. He told the population of the city after the massacre, 'For me the battlefield of France or Amritsar is the same.'[6] Nor in the subsequent inquiry did Dyer dispute that he had used excessive force; indeed that had been his intention. His aim was 'no longer a question of merely dispersing the crowd, but one of producing a sufficient moral effect, from a military point of view, not only on those who were present but more specifically throughout the Punjab'.[7]

Dyer was a soldier of limited ability, prone to hot-headedness. But the army backed him. Sir Havelock Hudson, the adjutant-general in India, defended Dyer's actions as militarily justified, particularly in view of the small size of Dyer's force and the large numbers of the crowd. In secret evidence to the subsequent inquiry Hudson emphasized the poor state of the British troops in India: undermanned, ill-equipped, and poorly trained, they would have been ill-prepared for sustained duties if Dyer's action had not contained the troubles of the Punjab. Monro, the commander-in-chief, tried to block the inquiry itself, arguing that it would call into question the army's authority. When it became clear that its report would censure Dyer for using excessive force, Monro pre-empted the government of India by relieving Dyer of his command and sending him home on sick leave.

The response of the army in India to Dyer's predicament was not only that of a profession closing ranks; it was too the product of the military's primacy in Britain's rule of the subcontinent. The response of the army in Britain, while shaped by these forces, was also the product of political conditions peculiar to home.

Dyer felt he had been condemned without a fair trial. This was a view shared by Sir Henry Wilson, the chief of the imperial general staff. However, the cabinet, having no wish to inflame the situation through further public debate, was anxious to dismiss him without a court of inquiry. Churchill, as secretary of state for war, told Wilson that the Army Council should endorse the cabinet's decision. But Wilson's concern was not with the political desirability of dampening down the affair, but with the morale of the army and its officers. He reckoned that Amritsar was likely to be the first of many such incidents, and that it was vital that the soldiers on the spot continued to have faith in the support of their superiors. 'It appeared to one, listening, that the story was a very simple one. The Frocks have got India (as they have Ireland) into a filthy mess. On that the soldiers are called in to act. This is disapproved of by all the disloyal elements and the soldier is thrown to the winds.'[8]

Sir Henry Rawlinson, who was about to go out to India in succession to Monro as commander-in-chief, agreed with Wilson. Actual discussion on the Army Council was stalled until such moment as Wilson had to be out of the country, in Germany. In his absence, the deputy chief of the imperial general staff, Major-General Sir Charles Harington, prepared a case which listed twenty-four factors in Dyer's favour. However, the generals had been outflanked by the actions already taken in India; they huffed and puffed about the memory of the mutiny and 'the interests of the army generally', but in the end could do no more for Dyer than endorse the line taken by Monro.[9]

The tension between the generals and Churchill became known to the press—unsurprisingly given Wilson's involvement. That supporter of the generals and scourge of the Lloyd George coalition, H. A. Gwynne, appealed on Dyer's behalf through the pages of the *Morning Post*. The massive sum of £26,317—approaching half a million pounds in 1995 values—was raised for Dyer. And, as a further throwback to earlier battles, the House of Lords passed a motion deploring the conduct of Dyer's case, and citing it as a precedent dangerous to the maintenance of law and order in the face of rebellion.

The inquiry into the Amritsar massacre had marked the limits of military responsibility in the case of civil disorder. It had concluded that Dyer did not confront rebellion, as he and his supporters insisted, but unlawful assembly. Furthermore, he had violated the principle of minimum force.

The example was a salutary one. The inter-war army came to argue that military force was a weapon of discrimination and restraint. It did so not just because of Dyer's fate. Its primacy in imperial policing was threatened by the advent of air power. The RAF argued that the bomber was a cheaper method of maintaining order; the army said it was indiscriminate.

Throughout the 1920s and 1930s, Palestine was largely administered by soldiers, Plumer being appointed high commissioner in 1925 and Lieutenant-General Sir Arthur Wauchope in 1931. But both proved reluctant to use punitive measures. The evidence that they were military men was largely restricted to their antipathy for the RAF; their administrations in other respects were marked by an excessive sensitivity towards the proprieties of civilian government. Only in 1938 did Orde Wingate's special night squads take the offensive against the Arabs, fighting terror with terror.

Minimum force was also a principal message of one of the two major studies of counter-insurgency operations published between the wars, Major-General Sir Charles Gwynn's *Imperial Policing* which appeared in 1934. Gwynn recognized the importance of co-operation between military and civil authorities. He saw three different conditions under which the army could be called upon to act. In two of them the army did not have autonomy—either because, although giving effect to emergency legislation, it was still subordinate to civil authority, or because the army was reinforcing the police. The third condition, and Gwynn's preferred one, was a declaration of martial law because then 'unity of control makes the military authority the sole responsible agent for carrying out the policy of the Government'.[10] But he did not develop the political aspects of his subject in the way that either Ireland (which he omitted) or Palestine (which he considered for 1929 only) suggested was necessary. His focus was on the tactics of riot control, and his approach to strategy was to insist that the making of policy should be left to the civilians. He failed to note that the absence of policy prevented the formulation of sensible tactics in Ireland, and was—even as he wrote—having similarly deleterious effects in Palestine.

H. J. Simson's *British Rule, and Rebellion* was written in response to events in Palestine in 1936–7. It was much more sophisticated in its handling of the subject, drawing a distinction between rioting and insurgency (which Simson called 'sub-war'). Policing was appropriate to the former but was too defensive and reactive to be effective against the latter. Simson defined insurgency as a form of political warfare, requiring a political as well as a military response. Like Gwynn, Simson's preferred method was martial law. Unlike Gwynn, Simson disliked the army's subordination to civil control. He saw the civil power in Britain as too neutral, too predisposed to act the referee in its pursuit of conciliation: it tended to see its police and its armed forces 'as one dog in a dog-fight, no better, and perhaps no worse, than the other dog'. His solution was the fusion of civilian, policing, legal, and military authorities in one co-ordinated body. A committee might advise the civil authorities on the preliminary steps, but in the emergency itself supreme command should be handed to a soldier. Because that soldier's remit would necessarily incorporate a wide variety of non-military tasks, his staff would have to be expanded to encompass the requisite outside expertise. He warned against underestimating the complexity of what he proposed: 'It is easy just to say that instead of the military aiding the civil power, the weight is just placed on the other foot and the civil power aids the military. There are, however, all manner of rules for military aid to the civil power, but none yet exist for civil aid to a commander exercising undivided military control.'[11]

Simson's book was published three years after Gwynn's, in 1937. The latter rather than the former became a textbook, and it was still recommended reading in 1945.[12]

The handling of Palestine therefore presented continuities that spanned the Second World War. Another general, Sir Alan Cunningham, was appointed high commissioner in 1945. But, despite his army background, Cunningham was not predisposed to the use of greater force. He became the opponent of martial law and offensive action. Cunningham had been removed by Auchinleck from the command of 8th Army in North Africa. His successor in that job, Bernard Montgomery, was now chief of the imperial general staff. Montgomery had served in Palestine in 1938, and was convinced that the offensive had worked against the Arabs then and would therefore be effective against the Jews in 1946. Montgomery's professional reputation exceeded Cunningham's, and he knew how to use it: the cabinet agreed to martial law. Wingate's methods of state terrorism found fresh application under the direction of one of his Chindits, Bernard Fergusson.

Palestine, like Ireland in 1919–21, was an object lesson on how not to achieve civil–military co-operation. Political solutions were out of step with military measures. The army and the police were poorly co-ordinated, with the result that operations were not based on a proper understanding of the threat. Propaganda was reactive, limited, and poorly focused. In the latter stages of the campaign, in the absence of clear policy, tactical imperatives usurped political direction and became ends in themselves. The effect of the state's employment of wartime veterans was a failure to distinguish between war and insurgency: the civilian population, treated as a potential enemy, was alienated rather than wooed. The principles of good policing were forfeit to the instincts of the soldier.[13]

In India in 1919 and again in Palestine in 1946, the army had tried to direct policy and in both instances it had failed. It had done so because it had been insufficiently politicized. It had seen the situation in military terms: in looking for an enemy it had gone some way towards creating one. What it had not so far done was to recognize fully that it could itself be the agent of policy in a much more active sense—that its tools need not necessarily be restricted to those of force. The exceptions to those strictures were of course the generals who had—in the manner consonant with colonial administration in the nineteenth century—been given political jobs. But the effect of these appointments—of Plumer, Wauchope, and Cunningham—had been less to educate the army in the relevance of civil–military fusion in counter-insurgency, and more to drive a wedge between these erstwhile soldiers and their former profession. Malaya, and above all what people believed happened in Malaya, changed all this.

In 1948 Sir Henry Gurney was appointed high commissioner in Malaya. Previously chief secretary in Palestine, he was possessed of convictions that stressed the demarcation between civil and military responsibilities, not their fusion. He believed that the withdrawal of civil power and its substitution by the army represented a victory for the terrorist. But in 1950 he proposed the selection of 'an experienced military officer for a new *civil* post', whose remit would be the preparation of a general plan of counter-insurgency.[14] Sir Harold Briggs, a retired lieutenant-general who had served on India's frontiers in the 1930s, was therefore given the task of co-ordinating civil and military operations. The 'Briggs plan' embraced the full spectrum of action in Malaya. It began with the premiss that the main support of the insurgent was the civil population. It set out to break this link through a 'hearts and minds' campaign, and through the resettlement of the Chinese peasant population in new villages, secure from the jungle in which the communists operated. Briggs developed the machinery of civil, military, and police co-ordination through a series of committees which operated at all levels of administration, not just the top. Gurney, despite his original preconceptions, found himself presiding over—and supporting—a system where the distinctions between the army and the police were being eroded, where co-ordination was leading to fusion.

In October 1951 Gurney was killed. Briggs was ill and due to retire. Malcolm MacDonald, the commissioner-general for southeast Asia, wished to revert to a system of dual control. In this he was supported by the actions of the commissioner of police, a former soldier whose service in Palestine encouraged him to develop a paramilitary force that competed in its functions with those of the army, and in doing so did more to divide the security forces than unite them. Oliver Lyttelton, the secretary of state for the colonies in the incoming Conservative government, took the opposite view. He argued that 'no line could be drawn to show where politics, civil administration, police action, administration of justice, and the like end, and where para-military or military operations begin'.[15] His solution was to appoint a civil–military supremo with joint responsibility, and to give that job to a soldier. In 1952 General Sir Gerald Templer went to Malaya as high commissioner and director of operations.

The constitutionalists, among them Malcolm MacDonald,

warned darkly of an impending military dictatorship. Dr Victor Purcell, a Chinese scholar and former army officer, said of Templer that 'with the powers of a Cromwell at his disposal, he often looked like the Lord Protector'.[16] Templer, his critics averred, created a police state that divided the entire population into soldiers or bandits, and that elevated the means of counter-insurgency over the political end.

All this exaggerated both the innovatory aspects of the appointment and the scope of Templer's powers. The need for civil–military co-ordination in counter-insurgency operations was hardly a new idea. It was the obverse of Lyautey's theories on colonial conquest; it was the product of previous British experience. Furthermore, the political solutions proposed for Malaya, above all the decision to accord it independence, continued to flow through constitutional channels; they were not usurped by Templer.

Templer's admirers are equally guilty of misrepresentation. Malaya, notably in the hands of another Chindit, Sir Robert Thompson, in his book *Defeating Communist Insurgency* (1966), became the model counter-insurgency campaign. It was contrasted with American failure in Vietnam. Templer's methods became the blueprint for application elsewhere, regardless of geographical, economic, and political differences. In particular two chronologically simultaneous campaigns—those against the Mau Mau in Kenya and EOKA in Cyprus—were assimilated with Malaya to form a general theory. The practice was not as clear cut as this implied. In Kenya it was decided not to establish unitary control. In Cyprus Field Marshal Sir John Harding was appointed governor and commander of the security forces, but his own claim that 'he was, in short, Templer by a different name' has been disputed.[17]

Templer's combination of political and military authority was both exceptional and short-lived: the emergency did not formally end until 1960 but dual control was restored on Templer's departure in 1954. None the less, for the army the powers enjoyed by Templer became the norm (and remain the norm) for successful counter-insurgency. In 1967 Colonel Julian Paget, a veteran of Palestine and Kenya, took three case-studies—Malaya, Kenya, and Cyprus—as the evidential spine for his book, *Counter-Insurgency Campaigning.* Paget concluded with five 'essentials for counter-insurgency operations'.

The first of these was civil–military understanding, the second a joint command and control structure.

In elaborating on the importance of civil–military understanding, Paget cited Templer in support, and emphasized its applicability 'at all levels from the Cabinet to the Company Commander'. 'There is no purely military battlefield in counter-insurgency warfare', he went on, 'and the campaign can only be won by combined civil–military efforts.' Paget reckoned that such co-operation could be achieved by three methods. The first of these paid obeisance to the principle of overall political direction: the government must declare its national policy, and define the army's role in the execution of that policy. 'The second condition which is essential to effective civil–military co-operation is the establishment of a unified command, with the direction and control of the campaign in the hands of one person.' Paget cited as examples Harding and Templer—both soldiers. Thirdly, Paget advised, the civil and military organizations need to plan and prepare before any crisis arises. 'There must be the closest possible co-ordination between the Armed Forces, the Police, the Government Departments and the intelligence organizations, both civil and military, not only in the territories where insurgency is likely but also in Whitehall. Cold war campaigning and counter-insurgency both need to be studied, planned and controlled by an integrated team, in peacetime as well as when active operations start.'[18]

Paget's advice was pragmatic and sensible. But it pointed to a militarization of civilian administration unimaginable to Clode or the advocates of dual control. Paget's aim, and indeed the army's collective aim, was not the acquisition of political skills or political functions: it was the achievement of success in types of operations not amenable to measurement by conventional military yardsticks. It was a professional necessity. The politicization of the soldier was a side-product—and to some extent an unavoidable one.

Walter Walker was one of the foremost exponents of anti-terrorist operations to emerge from the jungles of Malaya. His pedigree was that of the Indian army. His father had been a tea-planter in Assam but served in the South African and First World Wars; he himself followed his grandfather into the Gurkhas. His formative military experiences were not those of European warfare, but the maintenance of civil order in India and the conduct of operations on the north-west frontier. In 1942 he went to Burma and fought the Japanese with a toughness and ruthlessness that he would later apply to communist insurgents. Walker embodied the imperial legacy of the British army, and he was convinced that in the early 1950s this had more practical relevance to the army's immediate problems than did the nostrums of conventional or nuclear war in Europe. Walker established a jungle warfare school to perpetuate the skills learnt in fighting the Japanese, and in 1952 Templer asked him to write a manual on the conduct of anti-terrorist operations.[19]

In December 1962 revolt broke out in Brunei. Anti-colonialist in thrust, it was believed to enjoy the backing of Indonesia. The Indonesians disputed this, and, although they mounted their first raids in April 1963, their army was not directly involved until February 1964. The generation of confrontation was not a one-sided process: among its foremost advocates on the other side was Walter Walker. He urged Malaya to prepare for war.[20] Given the command of all three services in Borneo, he was made answerable to the commander-in-chief Far East, and—after its creation in 1963—to the Federation of Malaysia. He did not have political powers, nor did he have overall control of the police. However, his own experience as a brigade commander in Johore in the late 1950s had left him in no doubt of the importance of close co-operation between the army and the police. Walker's strategy was one of 'hearts and minds': heavy rains made priorities of flood relief, and of medical and engineering aid. Tactically, his emphasis was on rapid response, and he made extensive use of helicopters to this end. But his approach was not purely defensive. Walker believed in using his beloved Gurkhas and the Special Air Service, re-formed for jungle warfare in Malaya, to penetrate enemy territory and so pre-empt Indonesian attacks. Walker's critics feared that Walker was promoting the danger to promote himself. While others saw the operations as localized and defensive, and Britain's policy as one of de-escalation, Walker was convinced that the campaign would be lengthy and its implications far-reaching: he saw himself as a shield against communism for all Indo-China.[21]

Walker caused offence, and not just through his campaigning techniques. In the political in-fighting of the services, his manner was unsubtle. He was the first tri-service commander in the field: his employment of helicopters upset the RAF, and his rejection of assault ships alienated the navy. He fought his own service on the reduction of the Gurkhas. In 1967 he was due to be retired as a major-general. His appeal against this decision was not accepted by the army. But Walker had established powerful allies in the Labour party. Denis Healey, now minister of defence, when in opposition had visited Walker in Borneo; he had been particularly impressed by the latter's advocacy of joint command. Fred Mulley, the minister of aviation and himself to become minister of defence in due course, was another ally. Walker told Mulley of his impending retirement, and Mulley spoke to Healey. In September 1967 Healey told the Army Board that Walker's position should be reconsidered. Walker became general officer commanding Northern Command in the rank of lieutenant-general, and in 1969 was promoted full general as commander-in-chief in northern Europe.[22]

Walker's experience and reputation were as a fighting general in counter-insurgency operations outside Europe: it was these that had shaped his political instincts, however crude. He had now been given a supremely 'political' job, that of a senior NATO commander responsible for the sensitivities of many allies, and for preparing their troops for operations the like of which he himself had no direct experience. He had in the interim become disillusioned with the Labour government and its defence policies: he opposed the withdrawal from east of Suez, and he advocated conscription. He appealed to the chief of the general staff to broadcast to the nation on the role of the army. Rebuffed, he effectively took on this role himself. There was a touch of Wolseley in Walker.

Walker saw the geographical distinction between Borneo and the northern flank of NATO as strategically irrelevant. Both were in the front line of an impending war against the global threat of communism. What irked him was the complacency of the west, which he interpreted as the direct consequence of a conspiracy by NATO's political leaders. The latter, bent on the appeasement of the Soviet Union, were hiding the magnitude of the threat from their electorates. In these circumstances, the soldier's job was to prod the politicians, and, if the politicians failed to respond, to speak directly over their heads to the nation as a whole.

While in Oslo Walker co-operated with Tyne-Tees television in the making of a documentary. He saw it as 'a once-in-a-lifetime chance of allowing a "converted" television producer to go to his public with a hard-hitting story of NATO which might wake the public up and make people understand what NATO is about'.[23] Denis Healey found the film alarmist and militarist. He did not block it, but he did alert the supreme allied commander in Europe, General Goodpaster. Goodpaster raised objections on security grounds, although it had already been vetted for such lapses. Walker was convinced that he was being silenced for his political views.

Walker left the army in 1972. He was now free to speak directly to the nation, and he did so. Two books, *The Bear at the Back Door: The Soviet Threat to the West's Lifeline in Africa* (1978), and *The Next Domino?* (1980), incorporated his thoughts, albeit in a somewhat random fashion and in syntax that was at times equally idiosyncratic. He paraded his professional qualifications: each volume incorporated his full rank and decorations on the title-page, and added the imprimatur 'Former NATO Commander-in-Chief Allied Forces Northern Europe'.

Walker was among the most vociferous of those sounding the tocsin in the west in the 1970s—an alarmism that earned him ridicule from moderates (a ridicule which with the hindsight of the 1990s seems justified) but deep loyalty from those of a similar persuasion. The basic message remained the scale of Soviet ambitions. The experiences of counter-insurgency were conflated with those of a senior NATO commander to provide a geostrategic picture of global proportions. Walker's position confronted, on a much wider canvas, the classic dilemma of counter-insurgency operations—at what point do the means begin to undermine the end? Do state terrorism or emergency legislation or imprisonment without trial, each of them justifiable in operational terms, ultimately subvert the liberal democracies which they are employed to defend? Although ostensibly Walker's crusade had as its goal the preservation of the values of western society, it seemed at times that liberal democracy was not one of them.

Utilitarianism was fundamental to both his books. The focus of *The Bear at the Back Door* was southern Africa, that of *The Next*

Domino? was Pakistan. In both, regimes that were deemed illiberal by the west—either because of their domination by a racial minority or because of their subordination to the army—were vaunted because of their role in the war against communism.

In 1965, Ian Smith's Rhodesia declared itself unilaterally independent rather than accept Britain's insistence on majority—i.e. black—rule. At the time rumours abounded concerning the possibility that Britain would use its armed forces to intervene. This would have given counter-insurgency a new twist. First, there was the question of the military credentials of white Rhodesians. Many, including Smith himself, had fought with distinction in the world wars; Rhodesia had raised a squadron for the re-formed Special Air Service in the 1950s. Secondly, some of those British troops that would be involved could be subject to divided loyalties. They might have relations in Rhodesia. Rather more had had recent experience in Kenya protecting a white settler population against the Mau Mau. Soldiers would be required to act against their own inclinations in the name of constitutionalism. Commanding officers of some but not all units consulted their officers concerning their views on a possible intervention in Rhodesia. The chiefs of staff were alleged to have told the government that the armed forces should not be used in support of their Rhodesian policy, which was thus confined to diplomacy.[24]

A potential Curragh had been averted. But the ambivalence in the British army's view as to who was right persisted, and even intensified as the Smith regime conducted its own counter-insurgency operations against black guerrilla forces, themselves supported by the adjacent African states. In *The Bear at the Back Door* Walter Walker praised the Rhodesian army as the only organized force resisting Soviet imperialism. He described Peter Walls, the chairman of the Rhodesian Joint Operations Committee, as 'a real professional, a true and inspiring leader, a man of decision and action who radiates confidence'.[25] The British Labour government, and specifically its foreign secretary, David Owen, was castigated for undermining the efforts of Walls and his men. Significantly in 1977, Owen chose a field marshal, Lord Carver, as the power-broker most likely to be successful in bringing together the two warring sides. Carver's title for this job was 'resident commissioner', but, as one of his principal tasks was to create an integrated Rhodesian army, he chose to assert the British army's presence by wearing his field marshal's uniform. This affirmation of the army's proconsular role impressed neither the Rhodesians nor Walker.[26]

In *The Bear at the Back Door* Walker's support for illiberal regimes extended to Vorster's South Africa. The country was praised as a bastion of law and order in a confused and violent continent; its white population was described as a nation in itself; and its prime minister was congratulated for his pre-emptive action against communism.[27] *The Next Domino?* extended this support for regimes whose commitment to democracy or to the rule of law was in doubt to embrace the whole arc of southern central Asia from Turkey to Pakistan. In Turkey the coup of General Kenan Evren in September 1980 had prevented 'an imminent civil war and the collapse of a key NATO member—the last remaining bastion in the Balkans and the Middle East against Soviet domination of virtually the entire region'. Evren's tone with politicians was 'paternal . . . rather than threatening'; his intervention in domestic politics was thus justified by external circumstances.[28] However, the real focus of *The Next Domino?* was not Turkey but Pakistan, then also under military rule. Walker's visits to the subcontinent had taken him once again to the stamping grounds of his youth, and the hospitality he had received from the Pakistan army had reawakened memories of its lineal antecedent, the Indian army in which both Pakistan's generals and Walker himself had passed their youths. The divided loyalties which imperial legacies could generate, manifested in Rhodesia, were obvious also in Pakistan.

Walker therefore condoned rule by the army in other states. The crucial question in the context of the present study is this: where did his belief in the primacy of the external threat and the priorities that it imposed on a government lead him in regard to the governance of Britain?

By the time that Walker published his two books his disillusionment with the Labour party was complete. Even Fred Mulley, to whom he was indebted for the furtherance of his own career, was written off as 'our Minister for Disarmament and Disgrace'.[29] James Callaghan's administration, dominated, as Walker saw it, by Marxists and trade unionists, was in the thrall of Moscow. The silent majority in Britain colluded in the 'chilling success of Marxist brainwashing', and as a consequence the 'present state of anarchy and subversion . . . amounts to industrial terrorism'. Britain was committing 'hara kiri'. Walker's presumptions were those of the radical right: a commitment to law and order, the revitalization of the police, the reintroduction of corporal and capital punishment, a halt to black immigration, and incentives to curb the emigration of white professionals.[30]

Walker's politics, it must be re-emphasized, were shaped by his experiences as a soldier of the empire. 'Hearts and minds' campaigns—of which his own in Borneo was an outstanding example—achieved their ends by assuaging political radicalism with social and economic betterment: this was in part how Britain had ruled India after the mutiny. At the same time the doctrine of counter-insurgency had elevated military imperatives to be the co-equals of civilian. Walker, like Paget, emphasized the primacy in counter-insurgency of 'unity, joint planning and joint operations'. But he went further: 'Joint operations entail control by a triumvirate—civilian, policeman, soldier—all under the single direction of a military Director of Operations.' The soldier was supreme.[31]

Walker did not go as far as the advocacy of military government for Britain. But nor did he condone the army's constitutional subordination. Given his belief that the Labour government of the late 1970s was imposing 'defence cuts that will bring about the dissolution of NATO' and cause the delivery of Europe to Russia 'on a plate',[32] he regarded it as the obligation of senior soldiers to speak out—to the government if it would listen, and to the public at large if it would not. Much of his ire for Lord Carver was the consequence of the latter's failure, as Walker saw it, to exercise his right when chief of the defence staff to protest directly to the prime minister about the state of Britain's armed forces. Walker himself wished to go further. He wanted the chiefs of staff to be required to give evidence to the House of Commons committee on defence, in order that their own views would enter the public domain, and so generate pressure on the government to respond to professional opinion. Furthermore, serving officers should be free to communicate to the press. 'The situation is now far too grave', Walker testified, no doubt in reflection of his own actions, 'for politicians to have a monopoly when it comes to making public statements on vital defence strategy. Senior serving "military" men can no longer afford to bow to the principle that they are prevented by formal boundaries from speaking their minds. Political etiquette must take a back seat.'[33]

Walker's practical remedy for what he saw to be inadequate funding for the forces was therefore similar to that of Wolseley— to give the professional head a voice in a political forum. But there was another parallel—that between Walker and Henry Wilson. Walker was blunter, more abrasive, and more outspoken than Wilson. Like Wilson, however, he was seen as a maverick within the army, not totally trusted by his service colleagues, and ultimately relying on political patronage for elevation to the professional pinnacle which he attained. Wilson got where he did thanks to Lloyd George; Walker thanks to Healey. It was therefore easy for the army to distance itself from Walker as it distanced itself from Wilson. However, as this book has argued, Wilson stood at the cutting edge of a bigger and more fundamental problem—the inherently political nature of a professional commitment. To marginalize Walker by emphasizing his undoubted idiosyncrasies would be as misleading as any comparable efforts to marginalize Wilson.

Once in high office, both Wilson and Walker fell out with their political patrons. A prime influence for Wilson in the early 1920s, as it was for Walker in the late 1970s, was the weakness of the government's response to trade unionism. Equally important for Wilson was Ireland. Walker, although not himself directly involved in Ulster, managed to embrace it within his global picture. He concluded that the Soviet Union was the driving force in the subversion of Northern Ireland, and castigated the government for the 'flabby way' in which operations there were being conducted.[34] Such frustrations were not exceptional: they were reflections of views widely held within the army as a whole, however nuanced, and even if expressed in more subtle ways or in less public places.

Ireland has provided a symmetry to the British army's experience of counter-insurgency. Having failed there in its first experience of such operations after the First World War, it returned in 1969, just as Britain completed its withdrawal from east of Suez. Thus it became the last campaign in a sequence of colonial disengagements. The army set out to apply there the principles which it had derived from its intervening experiences.

The implications, however, were radically different. First, Northern Ireland was not a colony, but an integral element of the United Kingdom. Secondly, the military–political equation which had created a sense of success in the earlier campaigns could not be made to work here. The army had 'defeated' the insurgents in a military sense in only a minority of its earlier campaigns. What had ensured 'victory' was timely political concession, most often resting on the abandonment of the country concerned. But in Ulster the majority of the population was against British withdrawal. There was no readily available political solution, and the army therefore had no clear political strategy around which it could frame its military measures. At times the army was itself the *de facto* government of Northern Ireland. The usual relationship was inverted, and the army's success itself became a measure of success in wider terms.

When troops were first deployed in Ulster, General Sir Ian Freeland became director of operations, and the work of the Royal Ulster Constabulary was subordinated to his chairmanship. But the army's role was defined as that of peacekeeping: it had no political directive, and indeed it was answerable to two political masters—the Stormont government in relation to its political role in the province, and the Westminster government in its function as one of the nation's three armed services.

The establishment of direct rule from Westminster in March 1972 simplified the army's position. Indeed some have seen the decision by Major-General Robert Ford, the commander of the land forces in Northern Ireland, to select 1st Battalion, the Parachute Regiment for the control of demonstrations in Londonderry on 30 January 1972, so-called 'Bloody Sunday', as a deliberate ploy to highlight the army's frustration over the lack of long-term political direction. The Paras had a reputation for toughness which was reflected in the shooting of forty-two demonstrators, of whom ultimately fourteen died.[35] The outcome expedited the move to direct rule, while worsening the chances of a political solution.

For the next five or so years the army became the dominant force in the government and administration of Northern Ireland. Formally, civil supremacy was sustained. Martial law was not imposed.

But the emergency legislation of 1973 and 1974 permitted internment without trial and suspended trials by jury. Under the Emergency Provisions Act of Northern Ireland of 1978 a soldier on duty could arrest without warrant and without explanation not only those who had committed an offence but also those whom he might suspect of being about to commit an offence. In the same vein, the army was vested with powers to search premises, and to stop and question individuals. If it came to violence, the rules of engagement directing the soldier, the so-called 'yellow card', were not a form of public guidance, which might therefore inhibit certain forms of dissident behaviour, but a secret document.[36]

The army used much of the legislation given it under emergency powers not for police work *per se* but for the gathering of intelligence. The army, after all, was ill-equipped to conduct criminal investigation. But its earlier counter-insurgency campaigns had convinced it that success rested on the quality of its intelligence, and that that depended on masses of low-grade information rather than on a small quantity of high-grade material. Between 1971 and 1985 the army searched premises on 323,000 occasions: this effort peaked in 1973, when 74,556 searches were conducted, and tailed off after 1979. Over the same period, the police conducted only 11,852 searches. Arrests by the army climaxed in 1976, when 8,321 were made, and also diminished after 1979—a year in which the army was still responsible for 54 per cent of the total.[37] The need for such a high profile was generated by the collapse of the Royal Ulster Constabulary's Special Branch, but its effect was to generate confrontation rather than avert it.

The army also ran covert intelligence operations. As an immediate reaction to 'Bloody Sunday', the army was ordered for a period of months to adopt a low profile in Catholic areas. In an effort to continue to gain information, British soldiers operated in plain clothes. The Special Air Service posted individuals and, from 1974, a full squadron to the province. In emulation of principles developed in Kenya, 'counter-gangs' were formed to take the offensive to the Provisional IRA.

The difficulty of applying techniques refined in the colonies was the responsiveness of the press and the sophistication of the IRA in exploiting this. 'Bloody Sunday' was an example: the army saw its very title as evidence of the success of IRA propaganda. Another was the army's instruction of the RUC in the use of in-depth interrogation techniques, which it had developed from the experience of its own members when imprisoned during the Korean war, and had applied in Cyprus and Aden. It was accused of torture, a charge sustained by subsequent legal inquiries.

Thus controls which had proved uncontroversial in earlier operations proved counter-productive in this. Between 1968 and 1985 forty-five television programmes on Northern Ireland were banned, censored, or postponed.[38] Such intervention smacked of an unsatisfactory compromise: it carried the message that civil liberties were threatened, while failing to provide the monopoly of public information that the security services ideally sought. There was a parallel with the implementation of emergency legislation, which similarly affronted the tenets of liberalism while falling short of full martial law. The army's response, derived in part from its past experience with 'hearts and minds' campaigns, was to take a more positive approach to the press. Training in its handling became more widespread. The army's senior press appointment, the directorship of public relations, was increasingly bestowed on high-flying officers.[39] Soldiers learnt to project their views and to manage journalists on their own account: these were political skills, albeit acquired for good professional reasons.

During the 1970s therefore there were few areas in the administration of Northern Ireland in which the army was not intimately involved. Formally speaking, its dominance was reduced after 1975 with the introduction by the Labour government of a policy of 'Ulsterization'. The campaign was treated as a matter of law and order, terrorism as a criminal activity, and the police therefore given primacy over the army. The strategy of internal security increasingly owed more to the nostrums of liberal democracy than to those of counter-insurgency developed in the colonies. But echoes of the past were slow to fade. Sir Timothy Creasey had served in Kenya, South Arabia, and Dhofar, and was described as possessing 'Malayan views' when he arrived as general officer commanding in 1977. In 1979 he tried to persuade the incoming Conservative prime minister, Margaret Thatcher, to reverse the balance by appointing a 'supremo'. He was unsuccessful but a civilian co-ordinator, Sir

Maurice Oldfield, did set about improving military–police co-operation.[40]

Nominally civilian government did not preclude the continuance of widespread military powers after 1975, all the more insidious in some people's eyes for lacking the clarity of definition which martial law might have bestowed. Ultimately the soldier who opened fire remained bound by common law: the 'yellow card' had no legal status. Thus he could be charged with murder if that force was not 'reasonable in the circumstances'. The army felt understandably frustrated by what it saw to be the fiction embodied in this position, the idea that it was not at war. However, the success of its management of the press enabled it, most obviously in the case of Private Lee Clegg of the Parachute Regiment, to moderate the full force of the law in regard to its own members.[41]

A pioneer in the application of many of the army's counter-insurgency techniques to Northern Ireland was Frank Kitson, who commanded 39 Brigade in Northern Ireland between 1970 and 1972. Kitson had served in Kenya, and his thinking on intelligence, on the role of counter-gangs, and on the importance of 'hearts and minds' was all fashioned by that experience. During the course of his Northern Ireland tour, the fruits of a year's defence fellowship at Oxford were published as *Low Intensity Operations: Subversion, Insurgency, Peace-keeping. Low Intensity Operations* began with an introduction by the chief of the general staff, General Sir Michael Carver, and concluded with the injunction that counter-insurgency should be given greater coverage in the syllabuses of the Royal Military Academy Sandhurst and of the Staff College. Much of the intervening matter was uncontroversial and would have been familiar to anybody who had read comparable works, such as Paget's. In particular, Kitson's remarks on the importance of civil–military co-ordination, and his openness as to whether this was achieved by a single commander or by a committee chaired by a civilian, were entirely consonant with earlier theory.

None the less, there were subtle differences. The instruction at Sandhurst, Kitson enjoined, should give 'particular reference to the political background'; at the Staff College 'the teaching should . . . concentrate on the problems of building up a campaign using *civil* and military methods' (emphasis added). Both were sensible injunctions. But they provided a striking contrast with the controversy generated at the Staff College by Henry Wilson in 1908. Wilson had urged his students to make certain presumptions concerning foreign policy. Kitson's advice concerned domestic politics. It was not hard to imagine such instruction fostering a predisposition to intervene, particularly when he reminded officers elsewhere in his book that one of their functions in the pursuit of civil–military co-ordination was 'to vet action proposed by other departments in pursuance of the government's long-term aims in order to ensure that it is not harmful to the operational effort'. The possibility of military means usurping political ends was thus implicit.[42]

In a similar vein, Kitson's discussion of the law and of emergency legislation owed more to his pursuit of utility than to any commitment to natural justice. He saw two alternatives—that the law 'should be used as just another weapon in the government's arsenal', or that it 'should remain impartial and administer the laws of the country without any direction from the government'. Kitson based his support of the second option as much on expedience, 'because it is more compatible with the government's aim of maintaining the allegiance of the population', as on a sense of morality. Furthermore, the second option, he went on, 'might also prove unworkable if it were found to be politically impossible to get sufficiently severe emergency regulations on to the statute book'.[43]

Kitson's pragmatism was particularly evident in his belief that revolution grew out of minor, non-violent action. For him strikes, sit-ins, street-corner meetings were all the first, non-violent phase of insurgency. To be properly prepared the army had to become involved in planning and co-ordination at this early stage: ideally it was then that the machinery for civil–military direction should be put into place, psychological operations (or 'psychops') should be begun, and the process of amassing low-grade intelligence commenced. The implications were fraught: the army and the government would move over to the methods of counter-insurgency before insurgency itself was evident. Kitson begged the question as to whether all strikes and sit-ins had the ulterior motives he ascribed to them. The possibility that the army might generate the very conflict it was trying to avert was brought home with chilling force by Kitson's own maxims: 'Fighting subversion or insurgency is no more of a special subject than is the fighting of conventional war. It is all part of the same subject, i.e. fighting, and the only rational way of approaching the problem is to teach it as a perfectly normal phase of war.'[44]

To a sceptical and critical mind, Kitson was treating a striker or a protestor not as a citizen venting his right to free speech but as a potential enemy. The controversy Kitson's book generated arose not just because he was commanding a brigade in Northern Ireland at the time of its publication. He did not claim any immediate relevance for it, because, in common with most of the army at the time, he thought that that particular commitment would be of short duration.[45] He was addressing the needs of the late 1970s, not the early 1970s. And he made it clear that for him one possible theatre of operations was mainland Britain:

There are other potential trouble spots within the United Kingdom which might involve the army in operations of a sort against political extremists who are prepared to resort to a considerable degree of violence to achieve their ends. It is difficult for the British with their traditions of stability to imagine disorders arising beyond the powers of the police to handle, but already there are indications that such a situation could arise, and this at a time of apparently unrivalled affluence ... If a genuine and serious grievance arose, such as might result from a significant drop in the standard of living, all those who now dissipate their protest over a wide variety of causes might concentrate their efforts and produce a situation which was beyond the power of the police to handle. Should this happen the army would be required to restore the position rapidly. Fumbling at this juncture might have grave consequences even to the extent of undermining confidence in the whole system of government.[46]

Northern Ireland was therefore a stepping-stone. Kitson discussed one extended scenario in his book: it described tactics in a heavily wooded area where the village population was susceptible to the communist enemy. The colonial roots of *Low Intensity Operations* were thus laid bare. But its principles were being brought home.

Kitson's instant status as the *bête noire* of the left rested on the argument that the methods the army had developed in Africa and Asia were about to be used in Britain. The army had learnt how to control and manipulate populations through psychological operations; it had developed methods of mass surveillance; and it had used special units in covert, undercover operations. In Belfast these skills were being applied and refined in a major city of the developed world, not in a backward, rural community. A French journalist, Roger Faligot, wrote of Kitson in 1980: 'A man has declared war on Europe; a free Europe; a special war. A man distinguished by his rich military experience; his political outlook on warfare; and by his militaristic conception of politics.'[47]

All this exaggerated Kitson's influence as much as it distorted his message. But the evidence for its corroboration seemed plentiful in the 1970s. Kitson's own career, unlike those of many of his predecessors who had published influential books while in the service, blossomed. He went from Ireland to Warminster, to command the School of Infantry; he became commandant of the Staff College, and concluded his service as commander-in-chief of the United Kingdom Land Forces. Within the army his deadpan manner, and the uncertainty which it generated in his auditors as to whether he was pulling their legs or was speaking with deadly seriousness, made him a far from representative general. But to outside observers his continued promotion suggested that the army had not rejected his teaching.

There were other straws in the wind. After the terrorist attack on the Israeli Olympic team at Munich airport in 1972, the Special Air Service set up a counter-revolutionary warfare team. Its brilliant success in 1980, when it stormed the Iranian embassy in London, disturbed some as much as it reassured others. An élite military organization, which made as great a feature of its secrecy and low profile as it did of its toughness and resilience, had switched from the jungles of south-east Asia and the barren wastes of the Middle East, to the policing of central London. Reports of the SAS's doings in Ireland suggested to some minds that ultimately it was accountable to nobody but itself. In March 1988 it pre-empted an IRA bomb attack in Gibraltar by shooting three terrorists in very public circumstances. What alarmed many was the assumption of powers of execution by armed servicemen in civilian clothes whose policy was to shoot on sight without warning: what corroborated those fears was the fact that the terrorists who died in Gibraltar were not armed at the time. Even the words of the SAS's regimental historian, although designed to kill speculation, served also to foster it. Having insisted that its members were 'profoundly conservative with a small "c"', he went on: 'But more than most they have come to appreciate the political content of military activity.'[48]

Nor was the evidence of the army's tasks in Britain confined only to its specialist units. In 1974 troops were deployed at Heathrow for the first time, in what was described as training for the possibility of terrorist action at the airport: the left saw it as practice for a coup. The use of the armed forces to sustain essential services increased significantly in the 1970s. They provided cover for dustmen, firemen, and ambulance drivers, each on more than one occasion. The military intervened in industrial disputes twelve times between 1970 and 1981, and were standing by on fifteen occasions between 1970 and 1983.[49] The army, it seemed, was a strike-breaker, inimical to union activity, and actively sustaining its capability to administer the internal fabric of the state if called upon to do so.

In a book published in 1977, Jack Woddis saw Northern Ireland as a training ground not only in counter-insurgency but also in anti-democratic methods. 'The British army is being terribly brainwashed', he wrote, 'and acquiring the harsh outlook of a repressive, counter-revolutionary, anti-working class and anti-democratic institution which looks on those who are demanding democratic and national rights as the enemy. This represents an extreme danger to the democratic aspirations of the British people as well.'[50]

Woddis saw two dangers. First, the army, particularly now that it was no longer recruited by conscription, was increasingly separated from the main stream of society, and could therefore become a compliant tool in the hands of a repressive and authoritarian government. He presented his second fear as an extension of the first, although in reality it was more likely as an alternative. The army could, he reckoned, play 'a role as an apparently independent arbiter by indulging in its own coup politics and compelling the government of the day to capitulate to its demands for strong action against the popular movement, or even to force the government out of office altogether'.[51]

The question 'Could it happen here?' had suddenly become topical. Discussion of the British army's role in politics briefly attracted an attention which it had not merited before, and which it has not sustained subsequently. Adam Roberts, later to become professor of international relations at Oxford but then at the London School of Economics, wrote an article in August 1977 on 'The British Armed Forces and Politics: A Historical Perspective'. Roberts saw the army's political involvement as 'a perennial problem in Britain as much as in other countries'. 'Maybe', he suggested, 'there was never a "Golden Age" of civil–military relations.'[52]

The army protested. 'There is a world of difference', wrote Henry Stanhope, a defence correspondent sympathetic to its position, in 1974, 'between soldiers being used by the civil authorities to defend democracy—which is what the Army has in mind—and soldiers operating on their own account.'[53] However, the distinction which Stanhope made, that between objectives, was the easy one. The problem was that the two aims required broadly similar means. And means had a habit of distorting ends, in defiance of Clausewitzian norms. The outgoing chief of the defence staff in 1985, Field Marshal Sir Edwin Bramall, summarized the situation with a neatness that seemed to present more ammunition to his doubters than his supporters:

You may say that [an aversion to intervention for the purposes of sustaining law and order] is a most extraordinary attitude for a member of a Force, who over the last 50 years has done it again and again in Hong Kong, Singapore, Calcutta, Jerusalem, Bahrain, Aden and Nairobi to name but a few, which for part of the last 10 years has virtually been the civil power in parts of Londonderry and Belfast and which, as a result of all this, is now attuned as never before, and as no other Army, to the law and order role in terms of equipment, training methods, tactical doctrine and, above all, experience.[54]

Bramall and others reminded their readers or listeners that the army had not intervened in support of law and order within Britain since 1919. They cited the anti-Vietnam war demonstration in Grosvenor Square in 1968 as evidence to support their point. Nine thousand police were deployed, but because the priority remained the maintenance of law and order the army was not called in. The army, and the police, emphasized that Northern Ireland was not the same as Britain—that in the former government did not rest on consent, and that in the latter (where it did) contingency plans for the use of the army in a police role (as opposed to a civil disaster or political terrorism) did not exist.[55]

A major feature of the army's experience of counter-insurgency was of course its sensitivity to the press. It emphasized that its aversion to any internal use—whether strike-breaking or supporting the police—was the fear of a loss in popularity. This struck deep historical chords. Before the Police Act of 1856, the military was the state's main buttress to law and order. The army might reasonably have pointed out that throughout the eighteenth and early nineteenth centuries it had intervened in aid of the civil power without the whiff of a coup. But it was not a memory that the army cherished: it had been accused of using excessive force at Peterloo in 1819, but inadequate force in the Bristol reform bill riots in 1831. The army had been vilified as an agent of repression. The rise of the police and of the empire had changed things. Distance had lent enchantment; its equation with the empire had raised the army in the public's estimation. The two world wars—for all that rather more of the population had had direct experience of soldiering than it might have bargained for—had done nothing to dent that reputation. The army did not intend to lose it now.

Nor did it. From the perspective of the 1990s, the fears of the 1970s and early 1980s looked rather ridiculous. But the fact that the army did not intervene should not prevent us asking counterfactual questions. After all, the abandonment of empire proved sufficiently traumatic for the armies of France and Portugal, to name but two, to prompt a coup which failed in the first case and a coup which succeeded in the second. Why did the British army, for whom empire had been a far more persistent and far more pervasive influence than it had been for either France or Portugal, not stage a coup? Why did it not apply the techniques of counter-insurgency in Britain itself?

The first answer to this question must be external to the army. The thesis of this book is that armies are inherently political institutions only restrained from intervention by the political environment in which they find themselves. What prompted much of the speculation in the 1970s was the sense that government was weak, and that anti-government forces—particularly student radicalism and trade union militancy—were strong. After 1974, Harold Wilson's administration, having struggled to establish a working majority, was then circumscribed by the need to honour its own socialist credentials. Economic weakness squeezed the defence estimates. And the pursuit of détente left the army unsure of its role: it had lost the empire but was still feeling its way towards a whole-hearted European commitment.

In the 1980s this changed. The rhetoric of the Conservatives, if nothing else, endorsed the army, and victory in the Falklands heightened its prestige. More importantly, the government confronted the trades unions and won. Student militancy declined. The police forces extended their ability to reinforce themselves across county boundaries, and so enhanced their capacity to tackle major demonstrations from their own resources; more of their officers were armed, were trained in the management of explosives, and were qualified to counter terrorism. The possibility of their needing army support receded. Although temperamentally the values of the army stood increasingly at odds with the consumerism and self-interest of its parent society, politically the door to intervention—if ever it had opened—was edged shut.

Secondly, for the British army, the loss of empire did not mean a loss of purpose. Timely political concession hammered a sense of military victory out of a process of retreat: there were no Dien Bien Phus in Malaya. The empire itself did not disappear with the finality that the decision to withdraw from east of Suez suggested. The Falklands war was only the most obvious indication of that. But Northern Ireland itself was the most persistent. And in the early 1980s the increased confrontation with the Soviet Union, fostered in the United States by Ronald Reagan and in Britain by Margaret Thatcher, gave renewed purpose to the army's deployment in north Germany. This was particularly the case as the impact of new technologies on conventional warfare put fresh emphasis on the capabilities of land forces. Then, as the end of the Cold War and the possibility of peace in Ireland threatened to remove the principal functions of the army, first the Gulf war of 1991 and subsequently civil war in Bosnia emerged to take their place.

These are external explanations for the army's reluctance to intervene. However, there are also reasons internal to the army itself.

The most important is the Robertsonian tradition. In the twentieth century, the imperial and counter-insurgency roles of the army, which have tended to foster its political awareness, have been balanced by its commitment to the continent of Europe. The two world wars and the anticipation of a third have given practical force to Clausewitzian norms. Their effect has, as the previous two chapters argued, been that of integration and fusion. Frank Kitson's own writings have borne testimony to this. The command of a division in Germany and of the United Kingdom Land Forces put a different gloss on his views of civil–military relations.

In 1977 he published his mature views on counter-insurgency, presented in autobiographical form. The preface to *Bunch of Five* replied to his critics:

The army can only intervene when constitutionally called upon by its legitimate political masters to support the civil authorities, and when those authorities are unable by themselves adequately to contain the unlawful and unconstitutional acts of the insurgents and subversives.

The army should never—and in my view in this country can never—act in support of the civil authorities against insurgency and subversion in any other than a lawful and constitutional way. Such an assertion, which I regard as fundamental and to be understood as it were on every page of this book, can surely not be a matter of political controversy in Parliament or among law-abiding citizens.[56]

Kitson went further. He now made a clearer distinction between counter-insurgency and other forms of military activity. Insurgents used force only as part of a wider spectrum of political, economic, and psychological measures. The response of government would be similar. Civil–military co-ordination remained central, but in establishing its machinery and framework the soldier was only the adviser, not the executor. Kitson accepted that the consequence would be that the level of violence employed by the counter-insurgents would always lag behind that of the insurgents; if it did not, popular support at home and international approval abroad would be forfeit. The penalty for such political realism would be a war of long duration.[57]

The effect for Kitson was to push his experience of counter-insurgency and his reflections on conventional, high-intensity operations in convergent directions. He argued that the political and military leadership needed to be even more closely integrated in the nuclear age than it had been in the Second World War. The combination of the speed with which events would unfold and the alliance obligations under which soldiers would operate required a common awareness between soldiers and politicians of each other's problems. Reflecting the lessons of counter-insurgency, he highlighted the importance of public relations in winning popular understanding of the army's actions and so easing the pressures on politicians. Reflecting the legacy of Robertson, he stressed the responsibility of the chiefs of staff for protecting the field commanders from political pressure. They needed simultaneously to retain the confidence of their military subordinates and of their political leaders. To do this, they had to possess a background of normal military knowledge, plus 'a wide understanding of world affairs, current politics and of the way in which the government machinery works'.[58]

A general staff mentality was the positive reason why the army's political awareness was channelled in directions that made for co-operation rather than confrontation. But the army's apparent political quiescence was also due to explanations that, while intrinsic rather than extrinsic, were more negative. Increasingly after 1945 its political energies were being directed not outwards but inwards—inwards towards the other armed services, and inwards towards its own component parts. These two aspects of its political activities, which in a wider sense had the effect of neutralizing the army, will form the focus of the final chapters of this book.

9

THE POLITICS OF THE REGIMENT

The selection and training of officers are in the hands of institutions whose responsibility is to the army as a whole. These are the formative stages at which men's professional attitudes begin to be moulded. But running parallel to the collective mission is a subculture that is largely regimental. Overtly uniform, subtly but deeply differentiated—such is the character of the process by which men (and now women) enter the service of the crown.

Potential officers are assessed, and accepted or rejected, by the army's Regular Commissions Board at Westbury. But this is not their first contact with the army. Before going to Westbury they are put in touch with their future regiments, many of them attending pre-RCB courses run by their adoptive units. A former officer of the Royal Tank Regiment has described the operation as 'a very inefficient and uneven regimental quota and sponsorship system'. It is based 'on fairly well concealed regimental criteria', which often seek 'to fathom an individual's regimental motivation and soundness' rather than to establish strict competence.[1]

Cadets therefore arrive at Sandhurst already committed to a particular regiment or corps. But, if they are good, they will find themselves wooed by the competing blandishments of other regiments. Directing staff at the academy therefore have an ambassadorial role as well as instructional one, and great care has to be taken to ensure equal representation between groups of regiments, often to the detriment of arms of the service not organized on a regimental basis. Failure at Sandhurst is rare; acceptability in principle has already been determined by the Regular Commissions Board. Thus the training itself can at times be seen to be of little consequence. Since the 1960s, the course's duration has shrunk from two years to effectively one; over the same period the academic training of other nations has expanded, often to become degree-bearing.

The army's view is that the real education of an officer begins when he gets to his regiment. And the latter's role remains pervasive thereafter. Although an officer of ability will be selected for the Staff College at the age of 31, his quality will not accelerate his advancement. He has to wait his turn to command his regiment, a rotation dominated almost entirely by regimental committees. After he has graduated beyond regimental command, he moves into jobs that lie outside the regimental framework. But it is at exactly this point that the mechanisms for training and professional development become desultory and ill-developed.[2]

Entry to the army, and advancement within it, therefore rest on what R. G. L. von Zugbach has dubbed 'horizontal differentiation'. Outwardly all officers begin their careers on the same footing. In reality the social cachet bestowed by the regimental structure means that the teeth arms (primarily the cavalry and the infantry) can select from a large field of candidates; the other arms cannot. A potential officer, in choosing his regiment or arm, matches 'the military social system . . . to his own social background'.[3]

Thus from the outset, the individual regiment is an enormously powerful influence in shaping the officer and in helping him make the transition from adolescence to adulthood, from school to profession, from civilian life to military life. Its role in the selection and training of the officer has been graphically highlighted by John Keegan, for long a lecturer at Sandhurst. Keegan has gone on to conclude that the regiment is a prime influence in explaining the army's political neutrality.[4] Michael Howard made similar points in a review article in *Encounter* in 1962.[5] Their arguments consist of two elements.

The first is redolent of the theme of dual control, the idea of checks and balances. Through the regiment's mess, through its distinctive traditions, through its paternalism, the officer's loyalty is directed towards the lesser institution (the regiment), not the greater (the army). Rivalry and competition between regiments then internalizes any inclination in the army's officer corps as a whole to act more cohesively. The focus on dress distinctions, military bands, and regimental customs diverts militarism in Britain into safe channels.

The second element is social. The regiment with its county title (in the case of the infantry) or equestrian associations (in the case of the cavalry) carries rural connotations. It may have a royal colonel, and thus also link the army to the court. So the careerist and ambitious officer of middle-class origins, who might otherwise be predisposed to political intervention, is instead assimilated into a class structure that apes the gentry. His personal aspirations become secondary to the needs of the small, self-contained society which he has joined. True, the careers of those in the non-teeth arms are determined not by the movements of the unit but by individual postings, and in doing so foster more easily both personal ambition and, often, overt militarism. The latter, it has been suggested, may be a consequence of a desire to compensate: the technologies and skills of the non-teeth arms are often closer to those of civilian life, and their training is more centralized and command structure more bureaucratic. But any politicizing effects that service in—say—the Royal Army Ordnance Corps or the Adjutant-General's Corps might encourage are negated by the fact that it is the teeth-arm regiments which enjoy—through 'horizontal differentiation'—the near-monopoly of power and prestige in the army and so come to dominate its upper ranks.[6]

This is as far as most discussions of the regiment go. Because the British army is seen as inherently apolitical, there is nothing more to be said. But if the discussion starts from a different perspective— from the expectation that armies will behave politically because of their nature as institutions, and that what prevents them doing so is the maturity of the political culture in the nations of which they are parts—then the obvious conclusion is somewhat different. The regiment does fragment the British army, but this therefore means that the regiment itself has the potential to become a focus for political activity. In considering the politics of the British army, the regiment must constitute an important line of inquiry.

The first thing is to be clear about the false historicity that is part of the regimental penumbra. In the case of many regiments, there is an invented tradition as pervasive as that associated with the notion of dual control and the 1689 settlement.

Until 1881 all cavalry and infantry regiments of the line were numbered sequentially according to seniority. Thus the regiments with the highest numbers were the most junior. The number of regiments expanded in time of war and contracted in time of peace. In the Seven Years War infantry regiments numbered from 71 to 124 appeared and then went. During the Napoleonic wars the number of infantry regiments totalled 135; by 1818 the highest number still on the establishment was 93. Therefore, only the most senior regiments—that is those with low numbers—have an unbroken tradition of continuous service. Regiments with high numbers popped in and out of the *Army List* according to need. Although they shared the same numbers as their predecessors on each creation, the regiments themselves were formed from scratch. Crucially disbandment, not amalgamation, was the method of accommodating peacetime reductions.

In this there was nothing to distinguish British practice from the practice of most other armies—nothing to give the regimental system a cachet that it did not enjoy elsewhere. What happened in Britain was that certain features common to regimental systems in general lasted longer than elsewhere.

The most important of these was the fact that a regiment was a business. Regiments were effectively the property of the colonel, managed for him by the regimental agent. An officer invested money in the business either by purchasing his commission or by raising men for the crown in return for rank. He supplemented his inadequate pay by the surpluses derived from the money provided by the state to equip and sustain his men. The regiment therefore retained an administrative significance that was independent of its tactical role. Sir John Fortescue saw in the proprietary regiment a bulwark against parliamentary parsimony.[7] It was also of course an excuse for peculation, corruption, and misappropriation.

Tensions between the needs of the regiment as a business and the efficient management of the army as a whole were already evident in the eighteenth century. In 1748, the duke of Cumberland, as captain-general of the army, wanted to break the 28th Regiment of Foot because it was inefficient, and keep the more effective but more junior 48th. Cumberland lost. He was up against the vested interests of officers who had paid more for commissions in senior regiments precisely because they were deemed less vulnerable to disbandment. His struggle reflected a more general problem of the Hanoverian monarchy. Between 1714 and 1766, the crown endeavoured to subordinate the notions of regimental proprietorship to its own executive control. It enjoyed some success in the case of the captains, so contributing to the professionalization of the officer corps. But progress at the level of the company was not reproduced at that of the regiment, where colonels retained their proprietary interests.[8]

The practice whereby the colonel derived his pay from the profits which he made on clothing his regiment was not abolished until 1854, and elements of the proprietary regiment continued even later into the second half of the nineteenth century. Raising for rank was adopted to meet the manpower demands of the Indian Mutiny in 1857. The purchase of commissions continued until 1871. One of the arguments advanced against the abolition of purchase was that it would destroy the regimental system. Men financed their retirements by selling out to those junior to them in the regiment, so enabling the latter's promotion. Career advancement therefore tended to be by seniority and within the same regiment—at least in peacetime. Purchase also consolidated the status of the more senior regiments: as they were least likely to be disbanded, they constituted safer investments.

The fact that some soldiers suggested the abolition of purchase in 1871 shows that by the third quarter of the nineteenth century there existed a group of reform-minded officers who were prepared to challenge the dominance of the regimental system. Wolseley and Roberts were chief among them. Both developed their careers outside their regiments. Wolseley spent only five years with the 80th Foot; Roberts joined the quartermaster-general's department of the Indian army as soon as he decently could. The passions generated by the 'rings' associated with their names arose not least because of the challenge that the 'rings' presented to the regimental system. Wolseley and Roberts appointed to their staffs the best men available, disregarding the strict order of rotation for foreign service. By virtue of continuous campaigning, the 'rings' were fashioned into a system of advancement which was independent of the pecking order of regimental promotion. Both officers were effectively creating a general staff ethos—a sense that the army had a collective identity—ahead of the actual establishment of that institution.

A key problem for Wolseley and Roberts was the adaptation of the regimental structure to meet the needs of imperial defence. Most of the army's infantry regiments were single battalions. They therefore had difficulty in sustaining recruitment when posted abroad, often for several years. Men were enlisted for service in a particular regiment rather than in the army as a whole. Drafting soldiers from a regiment at home to make up the strength of one serving overseas presented contractual problems that went beyond the self-interest of the home-based regiment. Enlistment for general service was applied between 1799 and 1816, under the exigencies of the Napoleonic wars, but—apart from a brief flirtation with the idea between 1867 and 1869—was discounted thereafter.

The solution adopted by Cardwell, abetted by Wolseley, was to link battalions. He retained the existing numbered regiments but paired them, so that the regiment at home could provide drafts for the regiment overseas. In 1872–3, the pairs were allocated to a district, and were attached to the two militia battalions of the same locality so that the four could constitute a brigade.

The Cardwell system soon got into difficulties. It had been set up at a time of comparative peace, when the number of battalions divided between home and foreign service balanced. But by the end of the 1870s Britain was simultaneously fighting major wars in Afghanistan and South Africa. Therefore more battalions were abroad than at home, and those battalions that were at home were not ready for service as they were under strength and under-trained, their best men having been taken as drafts for the battalions abroad.

In 1876 a committee was appointed to inquire into the militia and brigade depot system. This seemed innocuous enough, but Wolseley was inspector-general of the auxiliary forces and therefore a member of the committee. It was argued that he hijacked its work for his own ends. Such suspicions were confirmed in 1881. The article which Wolseley wrote in reply to Roberts's Mansion House attack on short service associated the regimental system with conservatism.

He characterized the army's resistance to change as a 'disease of chronic grumbling and whining pessimism'. The cries on behalf of the regimental system and its *esprit de corps* are 'shibboleths wherewith to conjure on all military questions, just as a cry of "No Popery", or "The Church in danger", has before now been used to excite the masses politically'.[9]

Most of the evidence presented to the 1876 committee came from the militia officers, for whom the geographical titles of regiments had greater resonance than the numbered titles of the regulars. To the militia county names represented the places where they actually lived and served. Although the regular regiments had been given county titles, mostly in 1780, they meant little: regiments did not recruit there and had no base there. The committee concluded that the way out of the dilemma generated by Cardwell's reforms was to continue along the road on which Cardwell had embarked—to create brigades with territorial titles, each formed of two regular battalions, two militia battalions and a depot battalion. Officers and men would belong to the brigade as a whole, and not to individual battalions. The numbered titles would go.[10]

What was at issue was a matter of priorities—the best organization of the army as an entity against the feelings and traditions of the numbered regiments. The leader of the opposition to change was, however, the commander-in-chief himself, the duke of Cambridge, and he was backed by the queen. The duke was motivated entirely by his love of the existing single-battalion numbered regiments. 'The sentiment of an army', he said, 'is such that it is more important that anything else, and therefore I would not destroy regiments.' He confessed that it was he who had pushed Cardwell into linking battalions rather than amalgamating them in 1872, so as to preserve the identity of the existing regiments, although he recognized that the system would create problems in practice. His solution to the strains generated by the Cardwell system was not to go forward to multi-battalion territorial regiments, but to go back—to unlink the linked battalions. Although guided by sentiment, he produced a rational defence of his position. The army was an imperial army. Colonial conflict meant that it was, to all intents and purposes, constantly mobilized. The idea of territorial basing in order to facilitate mobilization, as practised in France or Germany, was a Continental system, inappropriate to the needs of empire. In the same vein, the duke opposed local recruiting. The army was an agent in the formation of Britain and a sense of Britishness: its regiments should therefore each be a judicious mixture of English, Irish, and Scots.[11]

Between them, the duke and the queen constituted a combination sufficiently powerful to get a fresh committee of inquiry appointed. Chaired by Lord Airey, who had been quartermaster-general in the Crimea, it was packed with the duke's nominees—old generals of the regular army, calculated to oppose the 'modern major-generals' and also the militia. No doubt much to the duke of Cambridge's annoyance, many of the witnesses felt that the linked battalion system was neither fish nor fowl, pointing out that inducting a man into a home-based regiment and then shipping him off to serve overseas in another worked against the creation of *esprit de corps*. Officers declared that they were reluctant to transfer into their linked regiment, even if it meant promotion. Many of the committee's witnesses therefore concluded that the correct route was to go forward on the lines suggested by the 1876 committee, or at least to create integrated regiments of two regular battalions rather than have two linked single-battalion regiments. None the less, the Airey committee went in precisely the opposite direction to that suggested by the evidence put before it. Its recommendations supported the line now favoured by the duke of Cambridge, proposing the unlinking of battalions and the maintenance of single-battalion regiments. It gave no strategic or organizational reasons for this beyond the fact that the linked battalion system was weakening battalions at home. Its main thrust was the need to restore the *esprit de corps* of regiments. The report was replete with references to regimental feeling, but it made no attempt to establish what that meant—to give concreteness to what was otherwise an intangible.

While all this was going on, the Conservatives had lost the 1880 election, and Gladstone was back in power. The new secretary of state for war, Henry Childers, decided to disregard the recommendations of the Airey committee and to proceed with the introduction of the multi-battalion large territorial regiment. The opposition would now be concentrated in the House of Lords, as it had been when the Liberals set about the abolition of purchase. Gladstone therefore wanted to ennoble Wolseley, who had been appointed quartermaster-general, so that the government could have a military spokesman in the upper house. But the queen would not co-operate.[12] The Lords debated the regimental issue on three occasions. In addition to the duke of Cambridge, a number of senior generals contributed: they included Airey himself, Lord Strathnairn, who as Hugh Rose had distinguished himself in the Indian Mutiny, and Lord Chelmsford, who had gained less distinction facing the Zulus in 1879.

The emphasis of those who opposed change continued to be on the primacy of regimental feeling, and not on the pattern of imperial defence or the needs of the army as a whole. The earl of Galloway, who proposed a motion against 'novel (so-called) "territorial" titles, inasmuch as this proposed substitution is known to be viewed as subversive of *esprit de corps*', cited the views of non-commissioned officers and privates. A sergeant of the 69th Regiment of Foot had declared: 'We hate the 41st already because we are joined to them.' A private of the 73rd, due to be amalgamated with the 42nd Highlanders to form the Black Watch, was equally trenchant: 'I am a Londoner, and don't want to be a Highlander and to wear the kilt. I know my regiment was the second battalion of the 42nd; but that was many years ago, and we don't want to go back.'

The earl of Morley reminded his peers that what they were discussing was the means by which to get the most efficient distribution of troops across the empire. He said that he 'was far from undervaluing the importance of *esprit de corps* in the Army. It was a sentiment which he honoured and respected in the highest degree, and one to which as little violence as possible should be done. But respect for this sentiment might be carried a little too far, if it were allowed to prohibit any changes of organization which were, in other respects, for the good of the Army at large.'[13]

Lord Morley went on to ask why it was not possible for the new territorially titled regiments also to establish their own *esprit de corps*. This was a point made by many other advocates of the change. They were of course right. When people spoke of the regimental system in the twentieth century, they had in mind the 1881 reorganization. The county titles which they venerated had no greater antiquity than that. Those who championed the regiment in 1881 were opposed to the system to which its subsequent defenders cleaved with such ardour. By the same token, those who instituted the 1881 structure thought they were putting the needs of the army over those of the regiment. Ironically, what they achieved in carrying through reform—to which they bent all the devices for political influence and parliamentary lobbying that the army possessed—was the reinvigoration of the regimental system. At a time when the armies of Europe were organizing themselves around the much larger unit of the corps, Britain was committing itself to a structure that at its theoretical best (the big regiment) was a fifth or a sixth of the corps' size.

What the reformers were hoping to institute was a brigade structure under another name. But what they got was a reorganization which ensured that the regiment remained the focus of professional loyalties. Two factors shaped this outcome: one was the establishment of local links, and the other was the priority of imperial defence.

The territorial titles embedded the regimental system in the nation. The county regiment became part of a local identity—tangibly so through the militia and volunteer battalions which now carried the same name as the regular battalions fighting in the empire. The new system contributed to the popularization of the army which was so marked in the last quarter of the nineteenth century. And the most fêted of all Victorian soldiers was the kilted Highlander.

One of the devices by which the opponents of the 1881 changes had tried to mobilize popular feeling against the reform was the fear that by amalgamating two single-battalion regiments the total number of individual Highland regiments would be halved. The very first objection which the queen raised when she wrote to the duke of Cambridge in 1877 was the problem of tartans which such amalgamations would create: 'to direct the 42nd to wear the Cameron Tartan, or my own Cameron Highlanders to wear that of the Black Watch, would create the greatest dissatisfaction, and would be unmeaning.'[14]

One solution to this problem, put forward in 1880, was to clothe all the Highland regiments in the Royal Hunting tartan. It might have been even more sensible to have suggested that they all wore the Black Watch tartan, which was after all a government pattern, and one from which many other post-1746 setts, including those of the Gordon Highlanders and the Seaforth Highlanders, had been derived. But Victoria would have none of it. 'The Queen . . . can decidedly not approve the proposal made with reference to the Highland Regiments. These are, most of them representatives of old Clans, and even in the present proposals *new* clan names are suggested. To take from them their clan tartans would be a great mistake, and to place them all in the Royal Hunting Tartan, which is a sort of undress Royal Stewart, will not be at all appreciated by the Highlanders, nor considered at all advisable by the Queen.'[15]

Childers mixed his exasperation with a sardonic humour. 'The tartan question', he wrote to Lord Reay in February 1881, 'is one of the gravest character, far more important . . . than the maintenance of the Union with Ireland. All the thoughts of the War Office are concentrated on it, and patterns of tartans, past, present, and future, fill our rooms. We are neglecting the Transvaal and the Ashanti for the sake of well weighing the merits of a few more threads of red, green or white.'[16]

To get round the vexed question of tartans, two second battalions, hitherto localized in England, became the second battalions of Highland regiments. The 73rd, which had ceased to be a Highland regiment in 1809, was grafted onto the 42nd or Black Watch. More insensitive was the handling of the 75th, which became the second battalion of a more junior regiment, the 92nd or Gordon Highlanders. As there were 141 infantry battalions in 1881, the pairing of regiments created 70 new two-battalion regular regiments, leaving one over. The unit chosen to be the sole single-battalion regiment was the 79th or Cameron Highlanders. These devices meant that there were five kilted battalions before 1881, and five kilted regiments after 1881. In the process the total number of Highland battalions was increased from five to nine.

By pandering to the particularisms of Highland regiments and their popular following, the army laid up for itself a fearful legacy in organizational terms. Scottish recruits may have been the ideal, but their numbers were actually in decline in 1881. In 1830 Scots formed 10 per cent of the population in the United Kingdom, but constituted 13.5 per cent of the army. By 1870 Scots were 10.5 per cent of the population, but 8 per cent of the army.[17] The clearances and Highland depopulation meant that the bulk of such Scottish recruits as could be found came—almost certainly—from the central Lowlands. None the less, the regiments in trews were not treated with the favour meted out to their kilted compatriots, but on the principles established for the English county regiments. Four regiments in trews—the 26th and 90th, and the 71st and 74th—were lumped together to create two two-battalion regiments, the Cameronians and the Highland Light Infantry.[18]

For the soldier, as opposed to the queen or the public, what rooted the new regimental system was not localization. The Highland regiments could only make up their establishments by recruiting, as before 1881, in areas outside their declared districts. Regiments took their recruits where they could get them, and for all of them that was increasingly the slums of large cities, particularly those of England.[19]

It was the empire that made the regiment so important to the soldier. Most infantry battalions spent more than half their service overseas. The battalion became the soldier's home, a sort of extended family, whose commanding officers regularly used the vocabulary of paternalism to describe its beneficent influences. Residence in alien lands only intensified the sense that the regiment was the embodiment of home. Moreover, the battalion was a sensible size for colonial campaigning, for imperial garrisoning, and for movement by troopship. In this respect the duke of Cambridge was quite right: the larger units adopted as the key administrative building blocks of Continental armies did not necessarily fit British needs given the daily demands of colonial service. Administrative and tactical practice therefore meant that the single battalion remained the spiritual focus of the army. The big regiment—the brigade organization formally adopted in 1881—became in many respects a dead letter. Each of the regular battalions of the new county regiments retained an independent ethos, spurning as far as it could postings and promotions that might have fostered homogeneity. The old numbered single-battalion regiments survived as the regular battalions of the new territorial regiments.

The system was therefore continuously under strain. The problem of finding drafts had not been resolved, and became acute even in the South African war. It collapsed entirely in both world wars.

By 1916 casualty rates meant that battalions could no longer be reinforced by drafts from within their own regiments. Complaints flowed into the War Office as a result of men being drafted from one regiment to another, or from one division to another. But the adjutant-general, Sir Nevil Macready, insisted that a sensible reinforcement policy was impossible without the maximum flexibility across the army as a whole. In doing so, he reflected the same commitment to the needs of the army, as opposed to those of individual regiments, which had inspired the 1881 reformers: 'In a war in which the whole nation is engaged *esprit de corps* must be for the Army as a whole and for the nation, rather than for the smaller units whose names live in the history of the wars in the past.'[20]

Macready had to confront the fact, however, that the localization of regiments had now enhanced their political leverage, and enabled them to stake their identities even against the imperatives of a world war. His answer was to avoid the mixing of nationalities: Scotsmen should go to Scottish regiments. The duke of Cambridge's idea, that the army was an agent for British homogenization, was deliberately set to one side.

This scheme was all very well, but in the case of Ireland the numbers of Irish recruits could not keep pace with the number of Irish regiments. In December 1914 there were fourteen Irish battalions at the front, and by May 1916 there were 54. By November 1918 this figure had fallen to 32. Irish voluntary recruiting was disproportionately low: in 1914–15, 7.8 per cent of the male population of Ireland aged 15 to 49 enlisted, as opposed to 24.2 per cent in England and 26.9 per cent in Scotland.[21] Furthermore, in 1916, when mainland Britain adopted conscription, Ireland did not.

Sustaining the strengths of three Irish divisions was therefore a continual struggle. Its implications went far beyond those of manpower. The 16th Irish Division was the standard-bearer of Redmondite nationalism; the 36th Ulster Division was the embodiment of Carsonite Unionism. Macready wished to amalgamate the two divisions in order to create a composite Irish division, but neither

Redmond nor Carson would have any truck with the other. When the matter was debated in the Commons in October 1916, military efficiency was being pitted against differing political persuasions in Ireland. The Army Council decided to keep the two divisions separate, and filled them either with English battalions (deemed to be more acceptable to either party than each other) or battalions of Irish regular regiments serving in other divisions. The effect was doubly damaging to the army as a whole. The two divisions were frequently well below establishment while efforts were made to fill them, and the expedients that the War Office was driven to adopt as a result of political intransigence disrupted the organization of other units.[22]

Logically, the experience of the First World War should have destroyed the regimental system. The mechanics of the Cardwell/ Childers arrangements were redundant in 1914–18. But the war reinforced the emotional significance of the regiment in two ways.

First, all new battalions were grafted onto existing regiments. The army expanded from 161 regular infantry battalions in 1914 to a peak of 1,750 battalions. But the number of regiments remained the same; they just acquired more battalions, often exceeding 20 per regiment. Each of the battalions of a regiment had very different experiences, served in different theatres, and developed different characteristics. They often had little more in common with each other than their titles and their cap badges. But the preservation of those titles made the regiments themselves important to a much broader swathe of the people of Britain. Before 1914 the army recruited from a small segment of the total population, predominantly the urban unskilled worker paid by the day and vulnerable to cyclical unemployment. During the war, classes and localities which in the past had had no direct military experience acquired links with the army. Virtually every family in the country was related to somebody who had served in action with an infantry battalion. That battalion was most probably a territorial or service battalion, but it carried the same title as the regular battalions of the regiment. It drew on the regiment's history and traditions. Fortescue, laying down his pen on the conclusion of the thirteenth volume of his history of the army in 1930, observed that the Great War was the first occasion on which the army had expanded without raising new regiments. The effect of adding battalions to existing regiments 'was to propagate the old regimental spirit, inherited from the Old Army, to the widest extent'.[23]

Secondly, regiments were now much more closely tied to the national divisions within the United Kingdom. The adjutant-general's policy of putting nationals into appropriate regiments reinforced those regiments' national identities. The Scots Guards, of whom a minority were Scots in 1914 (42 per cent), were 63 per cent Scots by 1918.[24] And the buoyancy of Scottish recruiting meant it was particularly in relation to the Scottish regiments that this sense of national identity was felt.[25] The Irish problem effectively went away in 1921, with the creation of the Irish Free State and the disbandment of those Irish regiments that recruited in southern Ireland.

Because there was no increase in the number of regiments between 1914 and 1918, the post-war reductions had little impact on the regimental system of the infantry. Instead the number of battalions in each regiment fell. The position was very different in the cavalry.

The cavalry was still numbered sequentially on the pre-1881 pattern. It had not developed local or (in most cases) even national links. It had not expanded in the First World War, and therefore had no redundant battalions to lose. Finally, its tactical role in modern war was, to say the least, uncertain.

In 1921, therefore, plans were drawn up for the disbandment of the four most junior cavalry regiments. As the adjutant-general, H. J. Creedy, pointed out, this was what had been done after previous wars, and there seemed no reason to change policy now. The draft army order clearly stressed the needs of the army as a whole over those of the regiment. It recognized that the Royal Tank Corps was likely to expand, and it took on board the need for financial stringency. It concluded that 'the retention intact of the whole of the Cavalry' would 'impair the general efficiency of the Army'.[26]

The seniority of the objectors who then wrote into the War Office supported the contention that the British high command in the First World War was dominated by cavalrymen. Allenby, as colonel of the 5th Lancers, disputed that his regiment was the most junior. Numerically it was not, but it had been disbanded in 1799 as the result of a mutiny, and was only re-formed in 1861. French, the colonel of the 19th Hussars, had to accept that his regiment was one of the most junior but pleaded that it had had more experience of modern war than many more senior regiments. He concluded with a typically question-begging touch: 'Finally it appears to me that Hussar regiments are more adapted to conditions of modern war than any other description of cavalry.'[27] Haig, the colonel of the 17th Lancers, was more indirect, but let it be known that he felt that the identities of all the cavalry regiments should be preserved.

What stoked this fire into life was the growing realization that five more cavalry regiments would have to go. While only four were affected, most of the rest of the cavalry were sufficiently self-interested to be prepared to see them disappear. But if nine of a total of 31 cavalry regiments were disbanded, the challenge was to a minority sufficiently sizeable to constitute a collective concern.

Therefore, in January 1922 Lord Cavan suggested to the man he was about to succeed as chief of the imperial general staff, Henry Wilson, that units be amalgamated rather than disbanded. He said, in the emotive language characteristic of those who spoke about the regimental system, 'the chief argument in favour of the retention of the souls of all the British cavalry is that it will prove to that arm that the Army Council are at least doing their best to preserve them, and that the spirit and tradition of the distinguished regiments of Cavalry will be kept alive'.[28] Creedy was aghast, citing historical precedent in favour of disbandment, and Wilson supported the adjutant-general. Cavan was persuaded of the wisdom of their arguments.

But the secretary of state now seized on Cavan's proposal. It constituted a device with which he could answer his political and parliamentary critics. In vain did Creedy and others point out that, whereas disbandment would destroy the identities of only those regiments selected for extinction, amalgamation would change the characters of double that number—both those originally selected to go and those to whom they were instead linked. Thus were born the so-called 'vulgar fractions', the 3rd/6th Dragoon Guards (from 1928 the 3rd Carabiniers), the 4th/7th Dragoon Guards, the 5th/6th Dragoons (from 1927 the 5th Inniskilling Dragoon Guards), the 13th/18th Hussars, the 14th/20th Hussars, the 15th/19th Hussars, the 16th/5th Lancers, and the 17th/21st Lancers.

Cavan's proposal proved doubly significant. First, it suggested that senior army officers should pay obeisance to regimental sentiment. Secondly, it established the precedent of compromise in place of confrontation, of amalgamation in place of disbandment.

The scheme adopted for the cavalry represented a set-back for the general staff's principle that the collective needs of the army should take precedence over the individual needs of regiments. The staff requirement was for a set number of cavalry squadrons: entire regiments of three squadrons were neither administratively nor tactically necessary. Many younger cavalry officers, who felt that amalgamation robbed regiments of their identities as surely as did disbandment, were of a similar persuasion. They favoured the idea of a corps of cavalry structured round the squadron, on lines comparable with the Royal Artillery and its battery organization. Instead, the War Office had accepted, in the words of the acting director of military operations, Walter Kirke, 'the views of existing Commanding Officers, who are naturally "crusted" horse soldiers'. Kirke felt that the cavalry had made a great 'mistake in not accepting conversion to tanks'.[29]

The general staff's loss of nerve created an inbuilt block to mechanization. The British army responded to technological innovation by adapting existing regiments raised for different and increasingly obsolete tactical roles, not by creating fresh regiments to meet new tasks. Thus, rather than expand the Royal Tank Corps early in the 1920s, it kept the tank corps small, and spent much of the inter-war period locking horns with cavalry regiments as to when they were to lose their horses and be given tanks. Even when cavalry regiments were mechanized, they persisted in seeing tank tactics in terms adapted from the cavalry, not as having characteristics peculiar to the new weapons system.

The reassertion of the regiment was justified, at least in part, by the fact that between the wars the army became once again a servant of the empire. Colonial garrisoning revalidated the regimental system in the infantry as well as in the cavalry. The artillery, which had become 27.5 per cent of the army as a whole under the strains of European war, fell back to 15.9 per cent of the army by

1935. Over the same period, 1918 to 1935, the cavalry and infantry grew in proportion, from 1.56 per cent to 5.4 per cent of the army in the first case, and 49.2 per cent to 52.6 per cent in the second. The arms of the service organized on functional lines suffered by comparison with those structured on regimental lines. The fact that artillery took the greatest cuts in relative terms meant that it had slower rates of promotion, and the chances of gunners reaching high rank were therefore proportionately lower than those of infantry officers.[30]

If the First World War was a war run by cavalrymen, the Second World War was a war run by infantrymen. Montgomery, Wavell, Slim, Auchinleck, Alexander—all were infantrymen; the first in particular showed a prejudice against the elevation of cavalrymen sufficiently marked to be seen as a legacy of Great War soldiering.[31] But inter-war soldiers, especially those of a generation too young to have experience of 1914–18, were the victims of a narrow professional education. Promoted at a stultifying rate owing to the consequences of budgetary restraints, their experiences and expectations were set by the regiment. The effects of this restricted outlook were to find reflection in operational ineptitude in North Africa in 1941–2.[32]

Furthermore, when confronted once more with major war, the Cardwell system cracked anew. As in 1914–18, men had to be posted to under-strength units when they became available regardless of local or regimental links. The administrative needs of the army as a whole were again set at odds with the particularisms of its component parts. Nowhere was this more evident than in the 51st Highland Division. Its infantry component, originally formed from territorial army units of north-east Scotland, had been forced to surrender at Saint-Valéry by the Germans' sweep through France in 1940. Although its Highland character was forfeit, its Scottishness was not. Major-General D. N. Wimberley, the man most responsible for the division's reincarnation, insisted that its *esprit de corps* was dependent on sustaining its national identity. He applied Macready's principles with a zeal that led him to tell his men that, 'if wounded and separated from your own units, do not allow yourselves to get drafted to other battalions, but see that you come back to us'.[33] In September 1943 1,200 convalescents from two divisions, one of them being the 51st Highland Division, were ordered to join the 46th Division inland from Salerno. Three hundred refused to do so, and 191 of them persisted even when their corps commander admitted that they had been encouraged to believe that they would rejoin their old units. They were charged with mutiny. Lieutenant W. J. Howat of the 7th Battalion, the Argyll and Sutherland Highlanders, one of the officers entrusted with the defence of those mutineers who belonged to the Argylls, told the court martial: 'To these men . . . General Wimberley is their military god and the sign "HD" is the altar at which they worship. So much so, that to ask a soldier of the Highland Division to fight with another division is, in my mind, akin to asking a Hindu to worship Mohammed.'[34]

The Salerno mutiny may well have been triggered by factors other than Wimberley's principles, but at the same time it demonstrated the impossibility of clinging to a system of recruiting that owed its origins to the 1881 structures. By 1944 the frontiers not just between nations but even between services were overridden by manpower shortages. The 2nd Battalion, Scots Guards, reforming in the borders for duty in north-west Europe, took in 400 reinforcements from the RAF Regiment, who arrived at Hawick still attired in sky-blue uniforms and defiantly chanting 'We are the RAF, RAF, RAF!' Dispatched to the Low Countries the battalion became part of the Guards Armoured Division. Its companies were affiliated with armoured squadrons from the Welsh Guards to form tank/infantry battle groups, so prompting the suggestion that they be called Celtic battle groups—a proposal rejected for fear of the ire of Glaswegian Protestants.[35]

As the war progressed the triumph of functionalism over territorial identity was confirmed by the pattern of enlistment. The adjutant-general, Sir Ronald Adam, was a gunner and a radical: his thinking was shaped much more by the requirements of a mass citizen army than by the colonial legacy of regimental structures. Soldiers were first recruited into the general service corps and then subjected to psychological and aptitude testing in order to establish the arm of the service for which they were best suited.

Complaints about Adam's abandonment of regimental reinforcement had no appreciable impact on his superior, the chief of the imperial general staff, Alan Brooke.[36] Brooke was another gunner. His focus was on the army, and he thought in terms of its needs rather than in terms of the regiment's. He demonstrated how over a fifty-year period up until 1945 the army, primarily through the general staff, began to develop a more corporate outlook. From Wolseley to Montgomery, Britain's senior generals ensured that the regimental system was subordinated to the army rather than equated with it. Although the system generated internecine warfare, its effects remained internal, and were contained and managed by the army's high command. Henry Pownall expressed these sentiments in somewhat negative terms when ruminating in 1941 on the decline of the army between the two world wars: 'What we need is a tougher army . . . One whose traditions are not based on purely regimental history but on the history of the whole British Army; where the competition is in efficiency, not in games or pipe blowing and band concerts.'[37]

In 1945, as in 1918, reductions were accomplished without the loss of cap-badges. As before, the army had expanded by adding battalions to existing units, and contraction was carried out by disbanding battalions to leave double-battalion or even single-battalion regiments. Furthermore, the advent of the Cold War, the maintenance of conscription, and the combination of global and European commitments meant that by 1952 the army totalled 440,000 men. In the decade after the Second World War, the army was able to avoid hard choices.

Then, in 1957, Duncan Sandys began the restructuring of Britain's defence policy. Its pivot was nuclear deterrence, and its corollary was a cut in conventional forces and the abolition of conscription. The army's response in 1957 was totally different from that of 1881 or 1922. On those earlier occasions a core of thinking shaped by the needs of the army or the general staff was clearly evident. In 1957 that was not the case. The army's senior officers made the regimental system the essence of the army as a whole, and fought to defend it. By equating the needs of the army with the needs of the regiments, they made the regiments not just the subject of domestic politics, internal to the army, but of the army's politics defined in a much broader sense.

On 27 November 1956, the Army Council organized a dinner to

celebrate the fiftieth anniversary of the creation of the general staff. The queen was the principal guest and she took as the theme of her speech the strength and virtues of regimental feeling.[38] The speech was doubly significant. First, she had been used by the army in a pre-emptive strike against its principal opponent, the minister of defence, Duncan Sandys: once again the army was exploiting the royal connection for political effect. Secondly, an occasion which was designed to commemorate a collective institution was hijacked to celebrate an organization that had so often stood in the way of an *esprit d'armée* in general and of the growth of the staff in particular. What was especially striking was that this had presumably been done at the prompting of the general staff itself. The army's high command was going 'regimental'.

The key figure in effecting this change was Sir Gerald Templer, chief of the imperial general staff between 1956 and 1958. Templer's own regiment was the Royal Irish Fusiliers. Recruited from southern Ireland, it had been due to disband in 1921. A popular campaign to save the regiment was successful when the Royal Inniskilling Fusiliers agreed to give up one of its two regular battalions, so that each could continue as a single-battalion regiment. The fact that compromise had preserved the identity of his own regiment seems to have been an instructive example for Templer: certainly he was disposed to amalgamation rather than disbandment.

Secondly, Templer's reputation was forged less in the Second World War (he had been seriously wounded in Italy, so suffering a brief check to the momentum of his career), and more in Malaya. Counter-insurgency operations were best conducted by small units. As in the nineteenth century and between the wars, the empire gave the single-battalion regiment a tactical and administrative relevance it could not sustain in European warfare. Templer did not believe that the war for which he was preparing the army would be what he called 'global warfare'—a high-intensity struggle centred on Europe and the use of nuclear weapons. Instead the threat from communism would manifest itself in revolutionary and insurgency movements throughout the world. Malaya was the model; the battalion was king. In July 1958, in his last speech as chief of the imperial general staff, he told his Camberley audience that 'In the British Army today, happy and contented units are more important for the future of the Army than units highly trained in [higher] formation training.'[39]

As a result of the Sandys white paper of 1957, the size of the army settled at 180,000 men. The infantry was cut from 77 to 60 battalions, and the armoured regiments from 30 to 23. In implementing these reductions, Templer followed three principles. First, amalgamation was preferable to disbandment. Secondly, recruiting records were more important than seniority: thus infantry regiments from contiguous geographical areas were to be linked with the aim of boosting enlistment. Thirdly, cuts were to be deepest where individual identities would suffer least. The Royal Artillery suffered most of all, being reduced by a third. Of the seven regiments to go from the Royal Armoured Corps, three were cut by amalgamating regiments of the Royal Tank Regiment, so saving three cavalry regiments. Function was at a discount.

The army's own report on restructuring, prepared in 1956 by a committee chaired by the incoming deputy chief of the imperial general staff, Richard Hull, had suggested that the infantry be reorganized into nineteen regiments, each of three battalions, one battalion to serve at home, one in Germany, and one outside Europe. He saw three-battalion regiments as adaptable to further reduction or expansion without the 'heartburning' such changes generated for single-battalion regiments: 'We feel most strongly', his committee's report concluded, 'that this is one nettle which must be grasped with courage.'[40]

Hull's predecessor, Dudley Ward, although himself an infantryman, was of a similar persuasion. He suggested regiments of four or five battalions. This was the logical development from the organization of the infantry in regional brigades which had been adopted in 1946, and which reflected what Childers and Wolseley had hoped they would achieve in 1881.

Templer did not grasp the nettle. He left open the option of large regiments, but he did not make them mandatory. His way was to set the broad parameters, and then leave the regiments to sort out their own futures and own identities. Regiments' abilities to survive became dependent on their facility for lobbying the press and agitating in parliament. They were free to assume independent political personas. And the decision not to publish the Hull report, taken on security grounds, meant that the debate was not set against any overarching structure. The parliamentary discussion about the post-1957 army was in many respects a discussion about regimental identities and not a discussion about the role of the army as a whole.[41]

The amalgamation which generated most heat was that of the Royal Scots Fusiliers, which recruited in Ayrshire, and the Highland Light Infantry, which recruited in Glasgow. The logic behind this union was geographical; the illogicality, at least for a military historian, was the mixture of heavy infantry with light infantry. The real bone of contention was neither of these things; it was the kilt.

The Highland Light Infantry was part of the Highland Brigade and had recently been permitted to abandon trews for kilts. The Royal Scots Fusiliers was a Lowland regiment and wore trews. The colonels of the two regiments put their heads together and concluded that the new regiment would wear the kilt, but that it would be of Royal Scots Fusiliers' tartan. The Highland regiments would not accept this: the amalgamated regiment would be in the Lowland Brigade, and therefore could not wear the kilt. Both colonels then decided that they would rather their regiments were disbanded than have to compromise. A protest march was staged in Glasgow in favour of the Highland Light Infantry; questions were asked in the House of Commons. In the end it became clear that the younger officers of the two regiments were more concerned about their careers than about their uniforms. Both colonels were asked to resign, and the amalgamation proceeded, to produce the Royal Highland Fusiliers, wearing trews.[42]

Templer's 'hands-off' approach to regimental restructuring was not the only reason why the army's internal squabbles had wider ramifications. There was also the split between his handling of the problem and that advocated by Hull. Templer's outlook was shaped by extra-European contingencies and the experience of counter-insurgency; in these the infantry was dominant. Hull was a 17th/21st Lancer; armoured regiments were increasingly geared for high-intensity operations in northern Europe. The idea that the army of the 1960s was being divided into two regimental blocks—global versus German, counter-insurgent versus conventional, infantry versus armour—may be crude, but it is also instructive.

Certainly it glosses over significant overlaps. Among them was Hull's tour of duty as commander-in-chief Far East Land Forces between 1958 and 1961. In the latter year he was appointed chief of the imperial general staff. Before leaving Singapore, Hull, a veteran of the war against Germany and the holder of every major staff appointment the army offered, spoke to Walter Walker, the embodiment of a totally different British military inheritance—India, infantry, the war against Japan, and counter-insurgency. The two were already at odds. Hull wanted the Gurkhas confined to internal security duties in Singapore, and chose another brigade for the wider roles contained in Britain's membership of the South-East Asia Treaty Organization. Walker had none the less trained his brigade to fight in any country and in all types of operations. Hull's message to Walker on this occasion was even less palatable: he told him that defence cuts in Britain made the Gurkhas particularly vulnerable, and that their establishment was likely to be reduced from 10,000 to 4,000 men. Walker was outraged. Given his belief that the struggle with communism was being fought in the jungles of south-east Asia, any cuts in the Gurkhas seemed folly. Regimental loyalty and strategic calculation coincided. If he had any doubts, they were quelled by the advice of a senior officer in the ministry of defence: 'For God's sake, fight, fight to the bitter end—because if you don't fight, nobody else will.'[43]

Hull had spoken to Walker informally. In 1962 Walker visited Nepal to convey his official report on Gurkha activity; he too spoke informally, on this occasion to the king of Nepal. The American ambassador in Kathmandu supported Walker: the foreign exchange earned by the Gurkhas in British service was an important element in Nepal's stability. Almost as he spoke, the revolt in Brunei provided the counter-insurgent scenario for which Walker had been preparing. Diplomacy and strategy had thus vindicated regimental loyalty. Walker wrote a paper reporting his conversations in Nepal for circulation within the Brigade of Gurkhas. Its thrust was reflected in an article in the *Daily Telegraph* on 3 January 1963, and the document itself was circulated among MPs by the chairman of the council of Gurkha colonels. The latter included both of Hull's predecessors as chief of the imperial general staff, Templer, who was colonel of the 7th Gurkha Rifles, and Harding, colonel of the 6th. Whatever their loyalties to the Gurkhas and to the principles for which Walker was fighting, they could not do anything other than see Walker as guilty of disloyalty. Walker had used a private conversation in dealings which had compromised Britain with the Americans. Hull was particularly incensed. He considered the possibility of a court martial. In the event Walker was relieved of his job as major-general, Brigade of Gurkhas.

Walker made an enemy of Hull, and, as we have already seen, nearly sacrificed his career. But it is hard to avoid the conclusion that behind the public posturing were many who were privately delighted by Walker's action: officers in the ministry of defence were only too happy to egg him on, while remaining reluctant to put their own careers in jeopardy. Furthermore, the long-term outcome proved far more favourable to Walker than the official response to his action suggested was likely. Throughout the 1970s and 1980s, the Gurkhas, albeit in reduced form, remained in the British order of battle, although the strategic rationale which had justified Walker's devotion to his regiment had all but evaporated. The message was clear—soldiering on in silence did not save regiments.

What should have finished the Gurkhas was the decision of the Labour government to withdraw from east of Suez. This prompted further reductions in the army, with a target of fifty infantry battalions being set in 1967 and 1968. The method used to effect the cuts was once again to leave the army to sort itself out. Those brigades that had formed big regiments on the lines advocated by Hull and Ward simply disbanded a battalion. Those that had not done so either had to amalgamate single-battalion regiments or disband one. Some very ancient and distinguished units, including the Cameronians and the Durham Light Infantry, opted for the latter course. The commanding officer of the Durham Light Infantry wrote to the editor of *The Times* to point out that a regimental organization created in 1881 might not be best adapted to the needs of 1968. His common sense was rare: what irked him was the attitudes of the Scottish regiments.[44]

The most junior battalion in the Highland Brigade was the Argyll and Sutherland Highlanders. Since 1967 the regiment had been serving in Aden, the last phase in the withdrawal from eastern empire. Its commanding officer, Colin Mitchell, was himself the son of an Argyll, even if—like many officers in Highland regiments—he had been brought up in the south of England (in Purley). He had served in the office of the chief of defence staff. While there, he had tried to publish articles but had been foiled by army regulations. This had not prevented him from developing a flair for publicity, a sense of how to manage the press, and a declared contempt for party politics: when he left the army he became first a war correspondent and then a Conservative member of parliament.[45]

Mitchell knew that his regiment was under threat. He argued, as every regimental commanding officer naturally argued, that his regiment was particularly special. It was different because it was a Highland regiment, and it was different from the other Highland regiments. He made much of the fact that it was the 93rd (Sutherland) Highlanders to which W. H. Russell, *The Times*'s correspondent in the Crimea, was referring when he wrote of 'the thin red line' at the battle of Balaclava—an early instance of journalism advancing the cause of the Argylls.

In Aden, Mitchell made sure his regiment had a high profile. Its men replaced their helmets with glengarries, and the Highland Brigade cap badge—known to the Jocks as 'the crucified moose'—with the boar's head and wild cat, the massive and unmistakable insignia of the Argylls. Pictures of Mitchell and his men were rarely off the front page.

Mitchell undoubtedly overstepped the mark. His reluctance to implement the policy of minimum force, and in particular his action in re-entering Crater, outraged his superiors. He was the only regimental commanding officer in Aden not to be awarded the Distinguished Service Order. But even these set-backs he turned to the public advantage of his regiment. The press dubbed him 'Mad Mitch'. When the Labour MP for West Lothian, Tam Dalyell, raised in parliament the report that grenades had been found distributed to platoons of the Argylls in Aden, although orders required them to be kept at regimental headquarters, his intention was to question the regiment's discipline and its manipulation of its Aden record. He achieved the opposite: the effect was to give the Argylls more publicity, and confirm its image as a tough fighting unit.[46]

On its return from the Middle East in 1968, Mitchell took the entire battalion to Scotland. Parading in May with its mascot, the pony 'Cruachan', its profile remained high, and Mitchell claimed that his was the best recruited of the Highland regiments. The other Highland regiments disputed this. Already there was a feeling that the Argylls were rocking the boat: if they proved successful in saving themselves it would be at the expense of a regiment less adroit in its handling of the press.

In July the government announced that the Argylls were to be disbanded. The council of the colonels of the Highland regiments had decided that they would not absorb another amalgamation (the Seaforths and Camerons had formed the Queen's Own Highlanders in 1961): they did not want the Argylls and the Argylls did not want anybody else. The regiment had to fend for itself.

Mitchell's publicity machine now swung into action. He himself had come out of the army the previous month, and so was free to publish his memoirs in 1969. A campaign to 'save the Argylls' was launched at the end of July, with General Sir Gordon MacMillan of MacMillan as its chairman. Full-page advertisements appeared in the press. Stickers were fixed to car windows—a form of advertising that was still comparatively new. The centrepiece was a petition, circulated in public places, principally in Scotland but also elsewhere. The original target was three million signatures by the end of August. Realism then intruded: the aim became a million signatures and in December it was claimed that this had been reached.[47]

Punch published a piece entitled 'The Argylls' Revenge!' which began with a spoof leader from *The Times*: 'Something akin to the rule of Cromwell and his major-generals seems to be developing in Britain. The incredible Argylls, unwilling to acquiesce in their own extinction, have been infiltrating themselves by devious and determined means into the seats of power.'[48]

Like all good satire, the article rested on a grain of truth. In parliament the key spokesman for the Argylls was the Conservative member for Ayr, George Younger, a former National Service officer with the regiment who was himself to become—much later—minister of defence. The shadow secretary of state for Scotland, Michael Noble, told the press that his views on the disbandment of the Argylls were 'quite unprintable'. In the House of Commons he committed the Conservatives to saving the regiment.[49] A few days later the Central Council of the Scottish Conservative and Unionist Association called on all Scots to sign the Argylls' petition.[50]

The Conservatives had committed themselves to the Argylls' cause. But they were not alone. Jo Grimond led the Liberals in a protest against their disbandment.[51] Scottish Nationalists pointed out that 'the bitterest aspect of the affair' was that 'the decision whatever it should be, was not ours to take'.[52] Support for the Argylls within Scotland therefore spanned the party political spectrum. The regiment was transmogrified into a national treasure. *The Times* quoted a former commanding officer, Colonel David Boyle: 'Across Scotland from Grangemouth to Campbeltown people feel something for the Argylls. It is very much a family regiment, full of fathers and sons serving together. Every single Jock knows that if he does badly the woman next door at home will know he has done badly. But the gnomes of Whitehall don't want that sort of Army: they want a lot of mercenaries, a lot of numbers, with no allegiance anywhere.'[53]

The needs of good discipline in the army, of justice to other regiments and other loyalties, demanded that a vociferous minority should not get its way. But the Argylls did win through. They were reduced to company strength rather than disbanded. Then, in October 1971, the Conservatives, who had been returned to office the previous year, redeemed their pledge. The Argylls were reinstated as a full battalion in January 1972. The regiment has not been touched in any subsequent reorganization. In 1994 when the Highland regiments lost another battalion they did so through the amalgamation of the Gordons, a regiment senior to the Argylls, with the Queen's Own Highlanders, itself a product of the 1957 review. The acquisition of a political identity served the Argyll and Sutherland Highlanders well.

The effect of this hands-off approach to regimental organization was to leave a structure that had no system. By 1971 the infantry consisted of six large regiments, each of three battalions; eleven regiments each of one battalion but formed by the amalgamation of two regiments; and a further eleven single-battalion regiments that dated from before 1957. The reductions since Sandys had been absorbed more by the loss of battalions in big regiments than by the establishment of the big regiment as a uniform model.[54]

Some called for an end to it all, and—in a step redolent of the general staff's position in relation to the cavalry in 1921—suggested the formation of a corps of infantry. *The Times* took up the refrain in a leader which earned a rebuke from Templer.[55] What Templer could not see was that he had fashioned an army designed for colonial policing: this alone could justify the general staff's abdication of authority to the regiments. For the next twenty years, between 1969 and 1989, the army reorientated itself, preparing almost exclusively for a European war and for operations in the north German plain. The single-battalion regiment was ill-adapted to this sort of warfare. The regimental system lacked the flexibility for the functions for which the army had now to prepare itself.

This switch in strategic perspective did not mean that the division of opinion embodied in the clash between Hull and Walker disappeared. It persisted but in a different form. The officers who reached the top of the army in the 1970s and 1980s had won their spurs under Templer. Those who had sweltered in Africa, the Middle East, or south-east Asia had, unlike those in the Rhine army, seen real combat. Many of them, although by no means all (Frank Kitson for example proved an exception),[56] defended the regimental system. By contrast, middle-ranking officers, who had served their apprenticeship in Germany rather than Malaya, tended to become frustrated with it. Their criticisms embraced tactics, administration, and promotion.

Regiments were associated with one arm of the service, and therefore with a particular tactical function. This created an obstacle to the introduction of new equipment as a change in weapons system challenged the identity of the regiment. The example of the cavalry's response to mechanization has already been discussed. Later illustrations have been provided by the army's response to the third dimension.

The Parachute Regiment does not boast great antiquity: it was formed in 1942. Its origins were functional, but its subsequent record and reputation quickly established an *esprit de corps*. It is a clear illustration of the point that regimental feeling does not need to rest on old-established traditions. However, the original purpose of the regiment has gone: massive airdrops of the type used in the Second World War are now inconceivable and it is unlikely that even a battalion will parachute into action again. Today, therefore, the Parachute Regiment performs the duties of ordinary infantry, albeit of particularly high quality. What has happened is that the regiment's own *esprit de corps* has created a rationale that has out-lived the regiment's declared function. The combination of red berets and toughness has ensured a public profile high enough to sustain its place in the *Army List*.

The Army Air Corps shows the problem the other way round. The armoured attack helicopter, designed for vertical envelopment as well as lateral, was one of the major innovations in the opera-tional level of war in the 1970s and 1980s. But the Army Air Corps and the helicopter found themselves in a position analogous to that of the Royal Tank Corps and the tank in the 1920s and 1930s. The Army Air Corps represented a challenge not only to the Royal Air Force but also to armoured regiments. Within a restricted budget, it could only expand at the expense of existing interests, and as a new corps it lacked representation on the Army Board. New functions could not be readily assimilated in a decentralized system domi-nated by regimental loyalties.[57]

Arguably new equipment should not be associated with the regimental structure at all. At what point does an armoured per-sonnel carrier, particularly if it mounts a gun, become a tank? The former is an infantry vehicle, the latter not. When does a tank become a self-propelled gun? The first is the responsibility of an armoured regiment, the second of the Royal Artillery. Tanks and infantry increasingly need to co-operate at tactical levels lower than that of the battalion: the battle group is a mix of all arms down even to platoon level. The interoperability and interchangeability of equipment at subordinate levels of command has meant that the army's fighting structures have been put at odds with the army's peacetime administrative organization.[58]

In reality of course this distinction—between the battle group for war and the regiment for peace—has not been sustained as rigor-ously as the nomenclature suggests. The sophistication and diversity of specialisms required by new technology have meant that the numbers of those within a regiment but not of it—of soldiers of technical units attached to the regiment—have increased. The Royal Electrical and Mechanical Engineers has proved the principal illustration of this point. The army cannot move without the REME. It is the most dramatic example of how functionalism is at odds with the regimental structure of the army. In 1989 10 per cent of the army wore the cap badge of the Royal Electrical and Mechanical Engineers, more than that of any other formation. Only one-third of the army was organized in traditional regiments.[59]

The focus on regiments therefore neglects most of the army. Where this imbalance in representation most obviously affects the service is in its senior ranks. Only two field marshals in the thirty years up until 1989 had not been drawn from the infantry or armour, and they were both gunners. Between 1963 and 1978 28.8 per cent of all generals (from major-general to full general) came from four sources only—the Household Cavalry, the line cavalry, the Foot Guards, and the Royal Green Jackets. The grip of these four became progressively tighter with seniority; over the same period they accounted for 46.4 per cent of all full generals. Between 1968 and 1978 the entire executive committee of the Army Board was drawn from the teeth arms—three-quarters of them infantry and armour, the balance sappers and gunners. Nor has the picture changed since then.[60] In 1989, the infantry constituted 27 per cent of the army, but provided 80 per cent of the army's lieutenant-generals. In the combat arms, one lieutenant-colonel in six could reckon on promotion to brigadier; in the logistic corps the chances were one in twelve.[61] Those in the army's supporting services have never—or almost never—been promoted outside their own corps: they remain as specialists, whereas cavalrymen and infantrymen are seen as generalists. The bias against the technical arms begins with the Regular Commissions Board, which is looking for the skill of junior leadership at platoon level, and continues right up to the Army Board.

The effect of the infantry's domination of the army's senior ranks was seen in the restructuring consequent on the end of the Cold War. 'Options for Change' set the total size of the army at 116,000 men, a reduction of 40,000. The infantry was to number thirty-eight battalions, including two of Gurkhas. Many middle-ranking officers, conscious of their experience within

NATO, and encouraged by developments in training for higher command in the 1980s to think in terms of the army as much as of regiments, recognized the desirability of radical solutions.[62] In the Gulf War units had to be brought up to strength with 'honoured guests', so generating friction over regimental identities.[63] But when the Army Board discussed the issue it remained wedded to the single-battalion system. The chief of the general staff, Sir John Chapple, was a Gurkha; in 1992 he was succeeded by Sir Peter Inge, a Green Howard; the commander-in-chief of the United Kingdom Land Forces, Sir John Waters, was a product of the Gloucestershire Regiment. None was from a 'big' infantry regiment, let alone from one of the corps. It is hard not to conclude that—understandably—their own regimental loyalties predisposed them against the sort of fundamental restructuring favoured by many.

And so in 1991 the Templer approach to reductions was repeated. The army was left to sort out for itself which regiments were to go and whether they were to be amalgamated or disbanded.[64] This suited the government admirably. The secretary of state for defence, Tom King, did not have to deal with any coherent argument about the army's role—about its future contribution to British strategy or about its optimum size. Instead the army turned in on itself, regiments jockeying for position in order to establish which would survive. Press and parliamentary discussion followed the army's own agonizing. The more basic question—whether an army of 116,000 was the right size given Britain's likely commitments—was relegated to the margins.

As in 1957 and 1968, it was the Scottish regiments that proved most adept in using political and parliamentary skills. Of the Lowland regiments, the Royal Scots—the most senior infantry regiment in the army—were slated to amalgamate with the King's Own Scottish Borderers. In the Highland regiments, the Queen's Own Highlanders were to link with the Gordons.

In August 1991 the Keep Our Scottish Battalions campaign was launched. (Its initials—KOSB—were of course the same as those of one of the regiments affected.) The campaign chairman, Lieutenant-General Sir John MacMillan, had just retired from the army as general officer commanding Scotland. Although himself a

Gordon, it was perhaps more significant that he was the son of the chairman of the 'Save the Argylls' campaign. Many of the methods of the KOSB campaign—the car stickers, the lobbying of MPs, the press coverage—owed much to the Argylls' precedent. The services of a firm of public relations consultants, Citigate Scotland, were retained. The campaign's key feature, as in the case of the Argylls, was a petition: at first the organizers hoped for two million signatures, but in the end they claimed 800,000.[65]

The charge to which the Argylls had proved most sensitive was one of selfishness—that if the Argylls were not disbanded another regiment would have to go instead.[66] The KOSB campaign met this accusation in three ways. First, and most obviously, it dealt with the futures of four regiments, not one. Secondly, and most constructively, it asked whether 'the peace dividend' on which the army's cuts rested really existed. Thirdly, and more negatively, it singled out those battalions which it felt should go in preference to the Scottish regiments. It asked why the Gurkhas should retain two battalions, the Parachute Regiment three, and the King's Division six. The rationale for each attack differed, but the central justification for internecine warfare was the recruiting record of Scottish regiments.

Like the Argylls campaign, the KOSB publicity gained in stridency through its association with Scottish nationalism. The policy of channelling Scots into Scottish regiments, initiated in the First World War and symbolized in the Second by Wimberley's 51st Highland Division, now reaped its whirlwind. Those north of the border were convinced that those in Whitehall failed to appreciate how strong was the feeling in the various parts of Scotland for 'their regiment'. The evidence suggested that the Scots were right. Pressed to explain the government's policy, Archie Hamilton, the minister for the armed forces, managed to strike the wrong note with extraordinary consistency. He acknowledged that, if only the best recruited were retained, regiments would be concentrated in Scotland, north-east and north-west England, and 'probably none south of the line drawn across from the Wash'. He deemed such a result politically unacceptable. On the one hand 'We have been accused', he wrote to D. A. Robson, ' . . . of sacrificing Scottish lives in any future conflict in order to protect the English and Welsh'; on the other, he told the Select Committee on Defence, 'people in the south of England would resent it and resent it bitterly'.[67]

Hamilton was wrong. The people of southern England cared very little for the Royal Anglian Regiment and the Queen's Regiment. The creation of big regiments had gone further in the south and had resulted in a loosening of regional ties. This may or may not have been a cause of their levels of recruitment: the latter might also have been a consequence of low rates of regional unemployment. What Hamilton's argument certainly did confirm was that the creation of big regiments had given the army, as Hull and Ward had anticipated, a greater flexibility in meeting change.

Thus the KOSB campaigners began to suspect that the Army Board was not the natural ally it thought it ought to be. MacMillan argued that there should be enough infantry battalions to ensure intervals of twenty-four months between tours. The Army Board, relying on the end of the Cold War to bring stability, seemed to think this possible despite the proposed cuts. Moreover, for some the recently won focus on the defence of Europe had wrought an intellectual shift. The operational level of war would be conducted at corps level; if Britain wanted to be a serious player among world armies it needed to give its generals experience of leadership at that level. The only way of achieving this after 1991 was through NATO's Rapid Reaction Corps, the command of which Britain secured thanks to its contribution of two divisions, structured for the high tempo of manœuvre warfare. Thus the numbers of infantry battalions to cover tours in Northern Ireland and elsewhere had to be weighed against broader issues of doctrine. MacMillan put the infantry view to the chief of the general staff, Sir John Chapple, and to his successor, Sir Peter Inge. The campaign's organizers were probably wrong to see either officer as an opponent of the single-battalion regiment (after all both their own regiments came through comparatively unscathed), but they were right to recognize that opinions within the Army Board were divided on the issue, and also to acknowledge the feeling in some quarters that a lot of the noise was being made by old and retired officers while not being echoed by the serving regulars.[68]

Two consequences followed from the Army Board's apparent lack of enthusiasm for the case of the Scottish regiments. First, the army as a whole did not present a united front; the divisions, although not particularly evident in public, could be turned to account by the government's supporters. Mark Tennant, beleaguered on two counts—as the Conservative candidate for Dunfermline East and as the treasurer of the Scottish Conservative Party, defended the government from the KOSB campaign by observing that 'there is a very large element within the Army represented at senior positions in the Ministry of Defence who see the army cuts as an opportunity for attacking the regimental system'.[69] Secondly, and more seriously, the lack of Army Board endorsement made the principal and most objective plank of the KOSB campaign's case—the operational need for more infantry—look suspect.

Consequently, MacMillan found it increasingly difficult to insist that the campaign's primary message should continue to favour strategic argument over a more populist 'tartan' appeal. Much of the rank-and-file support for the KOSB campaign was parochial. Within it were subsidiary organizations, the Friends of the Gordon Highlanders and Operation Borderer, whose impetus was much more emotive and much more political.

The atmosphere of Scottish politics had in any case changed markedly from that of twenty years previously. Hamilton was not the only Conservative minister to make the mistake of equating the mood in the Tory heartlands with the mood elsewhere. Scottish Conservatism was weaker, nationalism more strident. For the vociferous nationalists the campaign to save Scottish regiments was itself a Tory conspiracy. Scottish regiments did not represent regional identities: 'They are merely regiments of an artificial British state recruited from that area called Scotland.'[70] This claim, however contentious in itself, carried a corollary that was undeniably true— that the Scottish regiments had forged their identities in the service of the United Kingdom. The disbandment of those regiments would be a blow to the union. It was therefore the Unionist party that was strongest in their support.[71]

In 1969 the Conservatives had taken up the cause of the Argylls. In 1991 too the strongest grassroots support for the KOSB campaign seems to have been among Scottish Tories. This created an obvious tension. In 1969 the target was a Labour government; in 1991 Conservatives were turned against each other. The electoral vulnerability of the party in Scotland could not but become an issue. Although the campaign was launched on an all-party plat-form, Citigate advised that 'the Government needs to be convinced that failure to heed the campaign will be unpopular electorally, particularly in Tory-held constituencies'.[72]

Fortuitously the regiments selected for amalgamation came from the south and north-east, areas more winnable by the Conservatives than the central belt and the west coast. The Labour stronghold of Glasgow, whose local regiment—the Royal Highland Fusiliers—was not affected, responded to the KOSB campaign in lukewarm fash-ion. By contrast, an opinion poll conducted at the outset of the campaign in Galloway and Nithsdale produced evidence of stronger feeling. This was not only a recruiting area for the King's Own Scottish Borderers, it was also the seat of the secretary of state for Scotland, Ian Lang. Lang's majority at the last election had been 1,234. Of those canvassed, 88 per cent said they wanted the regiment retained, and 50 per cent thought it would affect voting patterns in the area. A month later, a poll taken of Conservative voters only found that at the next election no more than 60 per cent of them would vote Conservative again, and 24 per cent said their decisions would be affected by the fate of the King's Own Scottish Borderers.[73]

In October the death of Alick Buchanan-Smith, Conservative M.P. for Kincardine and Deeside, created a by-election in the heart of the Gordons' recruiting area. Those committed to the preserva-tion of the Gordon Highlanders reckoned that the issue was worth 5,000 to 6,000 votes. They planned to put up their own candidate, who was likely to split the Conservative vote. With their electoral position in Scotland so weak, the Conservatives could not afford to lose another seat. Ian Lang encouraged hopes that the Gordons would be reprieved. The campaigners felt sufficiently reassured to withdraw their candidate, and to opt instead for a written guaran-tee from all the candidates to the effect that they would support the Gordons.[74]

The tactics in Kincardine and Deeside seemed to have been successful. The determination among Scottish Unionists to chal-lenge their party directly was reinforced by another gaffe from Archie Hamilton, who implied that the Scottish regiments had suffered disproportionately because the government's already weak position north of the border left them with less to lose there.[75] D. A. Robson, a key figure in the Friends of the Gordons, suggested that at the forthcoming general election the methods used in Kincardine and Deeside be reproduced nationally. The KOSB campaign committee rejected the idea of putting up regimental candidates, but Robson and W. D. Fairgrieve of Operation Borderer urged companies which funded the Conservatives to reconsider their position. Telephone lines were set up to advise voters as to the views of particular candidates on the issue of regimental amalgamations. A 'Tories for the Troops Campaign' was organized by an Edinburgh councillor. Like many Conservative candidates, he believed that his chances of success were being prejudiced by the government's policy on the army.[76]

In the event the Conservatives, although hardly triumphant in Scotland, did better in the election than they had dared hope. Furthermore, Malcolm Rifkind, a former secretary of state for Scotland, succeeded Tom King as minister of defence. Rifkind was believed to be sympathetic to the regiments' case. In May 1992, a false report that the Royal Scots and King's Own Scottish Borderers were not to be amalgamated was seen as a deliberate ploy to draw him out.[77] The conviction that the Army Board was the principal obstacle and that the best route forward was to play on local, sentimental, and political chords grew stronger. Sir John MacMillan wrote an open letter to the prime minster, emphasizing the damage being done to the Conservative party in Scotland.[78]

The hopes placed in Rifkind were not entirely misplaced, even if the Friends of the Gordons proved to have been deceived. In February 1993 the minister of defence halted two amalgamations. One did indeed affect Scotland: the decision to link the Royal Scots and the King's Own Scottish Borderers was reversed. The other, however, halted the merger of the Cheshire Regiment and the Staffordshire Regiment. For the Cheshires, Bosnia, as for the Argylls in Aden, had come at the right moment. The press, bewitched as ever by the regimental system, had focused on their work and on their commanding officer, rather than on the longer-running commitment of less glamorous units like the Royal Engineers. The Gordons and the Queen's Own Highlanders went ahead to form the Royal Highlanders: the new regiment adopted the Gordon tartan for its kilts, Mackenzie for its trews (representing the Seaforth Highlanders), and Cameron for the pipes and drums.

Dress distinctions are one code by which the army reveals in outward form its inner loyalties. When John MacMillan was general officer commanding Eastern District between 1982 and 1985 he cut a dashing figure in a mess kit of glengarry and Gordons' kilt. It has become the fashion for senior officers, even if discreetly, to temper the uniform of their army rank with an indication of their regimental origins. Sir John Chapple, chief of the general staff during 'Options for Change', wore as a field marshal a uniform that owed much to the 2nd Gurkha Rifles.[79] Looking at the army, one might be excused for concluding that the regiments have colonized the general staff.

Significantly, when in 1995 John Redwood challenged John Major for the leadership of the Conservative party, he included in his statement of aims an intention to halt the loss of famous regiments. Beyond a promise to save the royal yacht, he said no more about defence policy. Symbols substituted for substance. By then the real problem for the army was no longer how best to absorb cuts in manpower but how to recruit even to the exiguous establishment vouchsafed it. In 1993 the Scottish regiments had seen the decline in UK-wide enlistment rates as a renewed case for their own survival; they had attributed their own falling enlistments not to a general trend but to the exceptional circumstances created by their own impending amalgamations.[80] But by late 1995 this looked like a misdiagnosis. Recruiting problems were universal. There was talk of using Gurkhas to make up shortfalls in infantry battalions: the notion of drafting Nepalese into regiments with British regional titles showed once again how the system cut across the need for flexibility.[81]

At the start of the twentieth century Robertson pointed the army towards the creation of a common ethos through its general staff, and argued that its sense of professionalism might give it political leverage and at the same time an awareness of the proper limits on political interventionism. As the century drew to a close his vision had become blurred. The regiment was the focus of the army's political activity. Soldiers were all too easily lured into the defence of their regiments because they were the units around which most of their professional careers were focused. Moreover, they realized full well that their outward symbols—uniforms and bands—were the tools with which it was easy to arouse affection and support in the wider community. The army was in danger of having its own preoccupation with the regimental system being taken at face value, of allowing forms to become driving forces in themselves.

In the narrow sense, therefore, regiments have not in themselves been productive of political neutrality in the army. Indeed, the opposite is true: the regiment has become the main focus of the army's political activity. But in a wider sense both John Keegan and Michael Howard are right. The regimental system has prevented the army speaking with a collective political voice. On 16 July 1994 *The Times* published an article by Simon Jenkins under the headline, 'Army Loses OK Corral Bunfight', and a subsidiary heading said, 'The Services, victims of cuts this week, should take a lesson from the BBC on the art of lobbying'. Jenkins was alarmed by the passivity of the armed services in accepting cuts of 5 per cent in the defence budget. In the army's case, part of the problem is not that it cannot lobby, but that the regimental system has meant that it is lobbying against itself, and that that suits its political masters only too well.

10

The Politics of Inter-service Rivalry

In September 1903 Admiral Sir Jackie Fisher, then commander-in-chief at Portsmouth but to become first sea lord in the following year, was invited to Balmoral by Edward VII. The king's wish was that Fisher should serve on the Esher committee. It was not a choice that commended itself to the army: Fisher told the king that he wanted 'an Army and Navy Co-operative Society by which the Army would be a Reserve for the Navy—like Nelson had the 69th Regiment on board at the battle of Cape St Vincent'.[1]

Nor was the Esher committee Fisher's first brush with the navy's sister service. In 1882 he had served with the naval brigade in Egypt. The fleet had bombarded Alexandria six weeks before the army arrived in Egyptian waters. The army was therefore critical of the fleet's effort, while the navy felt that its contribution had been undervalued. The fact that many of its shells failed to explode increased Fisher's frustration, since the navy was dependent for its ordnance on Woolwich, which was administered by the army. His experiences made him deeply opposed to the practice of providing naval brigades for bush warfare. He thought that it took sailors from their proper jobs and needlessly protected the War Office from the consequences of its own parsimony. Wolseley in particular had incurred his ire, not only for his mismanagement of the transport and supply services in 1882, but also for the derogatory remarks he had written about the navy in the *Soldier's Pocketbook*. Fisher's principal message to Esher was that 'The military system is rotten to the very core! You want to begin *ab ovo*! The best of the Generals are even worse than the subalterns, because they are more hardened sinners!'[2]

The nub of Fisher's struggle with the army was the division of the defence estimates. Navies needed ships and were capital-intensive: the switch from sail to steam had made this point especially compelling. Armies needed men and were manpower-intensive: the firepower revolution of the nineteenth century had not prompted a trade-off between men and machinery. Because ships took time to build whereas men could be enlisted by beat of drum, Britain in the eighteenth century had cut its navy much less drastically than its army in periods of peace. Between 1713 and 1763 the navy commanded about three-fifths of defence expenditure, and in peacetime about two-thirds. In the nineteenth century, however, this balance shifted: the demands of imperial defence eroded the distinction between peace and war, and in the period 1816 to 1895 (with the exception of the Crimean war) the army claimed 58 per cent of defence spending.[3] In 1889 the navy fought back. The Naval Defence Act allocated £21.5 million over five years to the senior service. Naval spending thereafter was set more by the navy's perception of its own need, and less by financial orthodoxy. By the mid-1890s the navy had once again overhauled the army. Fisher's target in 1904 was an annual allocation of £23 million for the army and £37 million for the navy. 'Every penny unnecessarily spent on the Army', he would repeat, 'means 2 pennies taken from the Navy.'[4]

Although the battle between the services was for cash, it was fought by means of strategic argument. The army had grown as a result of the acquisition of empire. It had also responded to the advent of steam by stressing Britain's vulnerability to invasion, demanding fixed coastal defences and disposable troops at home. The navy's reply was to point out that a powerful fleet was the best security for home defence, and to argue that the task of the army resided neither in the empire nor at home but in the mounting of overseas expeditions. The army, as Fisher was wont to put it in another of his refrains, was 'a projectile fired by the navy'. In the decade before the First World War Fisher's efforts were bent in two directions—to rubbish the fear of invasion and its associated demand for conscription for home defence, and to argue that the army's role in the event of European war was not to send an expeditionary force to the continent but to act as an amphibious arm in support of the navy's strategy.[5]

This fight for resources was not played out solely in the discreet confines of Whitehall. Fisher used to contrast 'the Court Army' with 'the People's Navy'. His case for naval primacy therefore rested on a management of public relations and a manipulation of the press so extensive as to make nonsense of the idea that the navy was 'the silent service'. The 1889 Naval Defence Act was a direct consequence of a scare developed by W. T. Stead in the pages of *Pall Mall Gazette*, taken up by the City of London, and supported by *The Times*.[6] Fisher sedulously cultivated J. R. Thursfield of the latter paper, J. L. Garvin of the *Daily Telegraph* and later editor of the *Observer*, and—especially—Arnold White who wrote in a variety of more popular prints.

The army of the late nineteenth century was no slouch in handling the press. Roberts and Henry Wilson in particular replied in kind. Vehement cases for conscription and for a Continental commitment were made on their behalf by Repington and others. In the decade before the First World War inter-service rivalry marched to the beat of Britain's strategic priorities.[7]

The case for integration not only within the army itself but across the two services rested in large part on the need to separate bickering from common sense. Symbolic of this desire was the dominance in the Committee of Imperial Defence of the Royal Marines—soldiers who served at sea. In 1904 George Aston, who had been in the Admiralty and also attended the army's Staff College, was asked to be the CID's assistant secretary: in the event he was spirited back to Camberley, where Rawlinson wanted him to lecture on amphibious operations.[8] His place was taken by another Marine, Maurice Hankey, and he in his turn recruited from the same source. In 1936 he asked Leslie Hollis to be secretary of the joint planning committee of the chiefs of staff: he said that he wanted a Royal Marine because he felt a 'neutral' figure would keep the balance between the services.[9]

The fact that 'neutral' carries connotations of negativity is significant. The Royal Marines' influence did not stop the Committee of Imperial Defence becoming a victim of inter-service competition, its function often being that of a referee between opposing sides.

Fisher's enthusiasm for 'an Army and Navy Co-operative Society', evident when he served on the Esher committee, waned when the call for a ministry of defence came from the War Office. Fisher feared that in the hands of 'Napoleon B', as he dubbed Haldane, a combined strategy would mean the domination of the navy by the army. Parliament considered each service's estimates in isolation, and the army and navy therefore shaped their strategic views in the light of this implicit competition.

In 1918 the Royal Air Force was created as an independent arm, and in 1921 its autonomy was confirmed with the establishment of the air ministry. The advent of a third service compounded the rivalry and confirmed the urgency of integration. Britain needed a structure that on the one hand did not subordinate the RAF to the requirements of the older services, and on the other did not result in the neglect of the older services' needs for integrated air support. Although the chiefs of staff committee was established in 1923 in response to this desideratum, it failed to produce a satisfactory solution to the dilemma. The navy's battle for control of aerial resources over the sea continued throughout the inter-war period. The army's failure to secure sufficient air support played a large part in its defeats between 1939 and 1942.

The RAF fought to preserve its existence against the background of cuts in the defence budget. It did so through the formulation of independent strategies. In the realm of imperial defence, it argued that air power was a cheaper and more effective form of policing than were land forces. In 1929 it claimed that in India, if five to six squadrons were added to its strength, up to thirty infantry battalions and ten artillery batteries could be disbanded for a net saving of £2,000,000 a year. The army, fearful that its traditional roles in the Middle East and on the north-west frontier were being cut from under it, countered that air power was indiscriminate in its application and provocative in its policing.[10] As the possibility of a renewed Continental commitment grew, the RAF again endeavoured to seize the high ground, contending in this case that the bomber would always get through, and that it would therefore either deter an aggressor or terrorize him into rapid submission.

Strategic bombing did not achieve all that its advocates claimed for it. Equally, imperial policing needed a mix of air and land forces, the precise ratio being a product of local circumstances and the objectives in view. The important point in the present context is that the most appropriate solutions were forfeit to the stridency generated by each service's need to secure its slice of the defence budget. The fact that the RAF advocated a ministry of defence reflected not a faith in the virtues of co-ordination, but a belief that that was the best means to break the dominance of the Royal Navy.

If left to the RAF, the inter-war army would have been confined to home defence against air attack.[11] The army was prompted to retaliate in kind. 'It is now sufficient', Pownall wrote in 1933, 'for the A.M. [air ministry] to say that Black is Black for the W.O. to protest that it is White.'[12] Thus the chiefs of staff committee became locked in inter-service disputes. In 1926 the chiefs of staff had committed themselves to the view that 'the size of forces . . . was governed by the various conditions peculiar to each service, and was not and could not be arrived at by any calculations of the requirements of foreign policy'.[13] The best balance, not the best strategy, became the end. And in the views of the army and the navy the balance was being upset by the RAF. At the same time each service felt that its own chief was failing it when the needs of inter-service politics forced him to compromise.

These divisions persisted beyond the outbreak of war in 1939. The chairmanship of the chiefs of staff was vested in one of the service heads, not in an independent chairman. For each meeting he held two briefs—one prepared by his own service and one prepared by the secretary of the Committee of Imperial Defence. The latter was designed to give the broader picture. But during the Second World War the preparation of the second brief lapsed.[14] Churchill became the independent arbiter. He was always anxious for offensive action as soon as possible, and in the circumstances of 1941–2 this delivered him into the hands of the RAF and its campaign for strategic bombing. The army resented what it saw as an inner nexus between Churchill and the chief of the air staff, Portal.[15] The priority put on strategic bombing directed aircraft development away from types suitable for ground support, and so confirmed the RAF in its belief that the task of the army would be no more than that of Germany's occupation after it had been broken from the air.[16]

Brian Loring Villa has shown how these circumstances constituted part of the background to the disastrous raid on Dieppe in
1942. Recognizing their comparative powerlessness, the chiefs of
staff colluded in an operation which they felt could be contained for
fear of anything worse. Both the Royal Navy and the Royal Air
Force were guided by considerations peculiar to their own service in
giving tacit acquiescence to a scheme with which they were
unhappy. The higher strategy for combined operations therefore
fell from its proper sphere to its executive head, Mountbatten. Villa
comes to the astounding conclusion that the latter never received
any formal authorization for the raid.[17]

The position improved from 1942 onwards. In part this was a
reflection of alliance pressures: the British needed to get their
strategic act together the better to negotiate with the Americans.
But it was also the achievement of Alan Brooke, who became
chairman of the chiefs of staff in March 1942. Brooke emphasized
the need to keep all three services in balance, and so create an
overall strategy. At the next level down, he opposed the practice of
appointing three commanders-in-chief, one for each service, to a
theatre of operations. He wanted a single supreme commander,
even if that meant the subordination of officers of two of the services
to the authority of the third. Both the first sea lord and the chief of
the air staff opposed him, but crucially Churchill agreed.[18] An early
and outstanding beneficiary of this arrangement was Mountbatten,
who took up the South-East Asia Command in October 1943.

Brooke's chairmanship vindicated the principles of co-ordination. He could be as intolerant as any soldier of the service particularisms of the navy and—especially—of the air force. The BEF's
lack of air cover in 1940 was a searing experience which could
render his demands of the RAF 'unhappily peremptory and highly
critical'.[19] In the long run Brooke learned to moderate the excesses
of each service without alienating their advocates. But he did not
abandon the army in the process; indeed his mastery of inter-service
politics enabled him to promote the army's cause. Despite its
dominance of British war-making in 1914–18, the army had spent
the years 1918 to 1942 in the shadow of the other two services, and
in particular of the RAF. Only Sir George Milne, of Brooke's seven
predecessors as chairman of the chiefs of staff, had been a soldier.

Disillusioned and demoralized, the service was in danger of losing its voice in the formulation of British strategy. By exploiting the ideas of balance and co-ordination, Brooke restored the army's status.

Implicit in Brooke's management of the war was the idea of a truly integrated defence policy. But the 1946 white paper on defence organization rejected radicalism. It spurned an all-powerful ministry of defence and a tri-service general staff in favour of 'a guiding hand to formulate a unified defence policy for the three services'.[20] A minister of defence was appointed, but he lacked either the personality or the structure to enable him to manage the individual armed forces. For much of the next half-century Britain moved towards the solution rejected in 1946, but in gradual and incremental steps. As it did so, its path was dogged by inter-service competition.

Brooke's legacy to the politics of the army was its promotion through the machinery of inter-service balance. Montgomery, who succeeded Brooke as chief of the imperial general staff in 1946, applied—at least outwardly—the principles of his predecessor. Formally speaking he was an advocate of integration. He wrote to Mountbatten in 1947, urging him to use his influence with the prime minister to create a 'Chief of Staff, Armed Forces, who would be an independent Chairman of the Chiefs of Staff Committee'.[21] In 1955 he addressed the Royal United Service Institution on the need for a powerful ministry of defence. He attacked the duplication and waste generated by the distrust between the services, the self-contained nature of each, and the pressure on the chiefs of staff to fight their own corners. He recognized that the tasks of each service overlapped, and that scientific and technical innovation did not necessarily fit into the received framework of the existing structure of the armed forces. Like Brooke, Montgomery also paid lip service to the dominance of air power in modern war, stressing the importance of not subordinating the RAF to the two older services.[22]

Whereas Brooke behaved as though he believed what he said, Montgomery did not. Brooke's successor as chairman of the chiefs of staff was the chief of the air staff, Tedder. Montgomery knew that Tedder had intrigued to have him removed from the command of

21st Army Group in Normandy in 1944. He hated Tedder, and he did not trust the cleverness of Cunningham, the first sea lord. His advocacy of greater integration was in large part a consequence of the clashes between his personality and the existing structure. He was 'determined to have the biggest share of everything for the army', and his method of achieving it was to table a paper to that effect without previous circulation.[23] When opposed, his response was not to negotiate, but to hector, to bully, and, ultimately, to bypass his colleagues. The pursuit of balance, which in Brooke's hands had served not only the ideal of inter-service harmony but also of army self-interest, was abhorrent to Montgomery: he saw it as sitting on the fence.

Montgomery may have set back the cause of co-operation, but he proved more successful in his advocacy of the army's position than his methods might have warranted. His constant assertions that he was a simple soldier were of course nonsense. He may not have been an able politician, but he was quite capable of using his professional reputation to achieve wider objectives. He subverted, as we have already seen, the policy of the high commissioner in Palestine. Internationally, while still chief of the imperial general staff, he visited the United States and Canada, and, without the prime minister's authorization, proposed a trans-Atlantic pact against communism. Domestically, he believed in 1947 that the government would have to confront the workers; he told his former chief of staff, de Guingand, that 'If the thing goes sky-high, I may have to play a part: and would do so.'[24]

Montgomery was a professional soldier, and it was to the army above all that his political instincts were harnessed. Even if not his alone, his most striking achievement was the adoption of conscription by a Labour government in peacetime and at a stage—1947—when the pattern of post-war strategy was still uncertain.

Montgomery's reshaping of the post-war army, his so-called 'New Model Army', was built on the experiences of 1914 and 1939. His aim was to have a trained reserve that could be rapidly mobilized in the event of war—a mass citizen army that would not have to be improvised and that would not take years to form. Conscription was a manpower-dominated solution to defence. Although it had the backing of the other two services, and in particular of the RAF, its main advocates—and above all the advocates of a minimum period of eighteen months' service—were to be found within the War Office. Montgomery's lobbying on its behalf extended beyond the chiefs of staff to include Conservative members of parliament. Churchill was angered by what he saw as military intervention in politics.[25]

The convertibility crisis of July 1947 and the subsequent cuts in defence spending brought home to the other two services that the effect of a manpower-dominated defence policy was a squeeze on equipment. Moreover, although conscription might enable the army to mobilize and expand more quickly than in either world war, its response was likely still to be too slow in the circumstances of the nuclear age. The RAF emphasized that it could develop a capability that would be both massive and rapid. In December 1947, Tedder, following wartime thinking on strategic bombing, set a target of 375 bombers capable of carrying atomic weapons. The RAF's strategy was both offensive and defensive—a strike force that could also deter. Its effect would be to shoulder the army into the margins of national security once more. The latter's task would be restricted to the defence of Britain and its overseas territories, and yet—as the other two services were quick to point out—the risk of invasion was minimal.

The army had to find a strategy to reassert its claims on the defence budget. The Berlin crisis of August 1948 and the onset of the Cold War enabled it to do so. In 1946–7 it had rested its case for conscription on the idea that war with the Soviet Union would not occur for ten years: the creation of a large reserve was a reflection of that long-term target. By 1948, the long-term objectives were associated with weapons procurement and the Royal Air Force. The army's focus was now on the immediate challenges generated by communism.

One challenge was global. Although the granting of India's independence in 1947 reduced the strategic importance of the Middle East, Slim—who succeeded Montgomery as chief of the imperial general staff in November 1948—saw the region as a buffer against Soviet expansion southwards. The Middle East did not lose its primacy until 1950, and by then both Malaya and Korea validated the army's global conception of the Cold War. The defence of empire had found a successor, at least for the moment.

The other challenge was of course the defence of Europe. A fully fledged Continental commitment was slow to grip the army, but the strategy was espoused by that service in the face of opposition from the other two. What it enabled the army to do was to find a role within a conceptual framework set largely by the Royal Air Force. First, it could argue that the threat from the air required forward bases in Europe both to protect Britain from air attack and to enable strikes into eastern Europe. Secondly, a token British military commitment to the defence of western Europe could help invigorate the Western Union and above all boost the French. By 1950 the army was beginning to find the role that until 1989 would give it increasing leverage in inter-service lobbies.[26]

The main obstacle to the maximization of strategy—apart from the army's own predilection for bush warfare—was the advent of nuclear weapons. The atomic bombers of the RAF were Britain's main tools for offensive operations in Europe. The Global Strategy Paper of 1952, largely driven by Sir John Slessor, chief of the air staff, accepted that the central pillar of Britain's strategy would be its nuclear bombers.[27] On the army's and navy's insistence, the paper acknowledged that there might be a period of 'broken-backed war', with fighting continuing at a conventional level after an initial nuclear exchange, but for the army at any rate its role in Europe was clearly a subordinate one.

The army's position was rendered even more vulnerable because of its reliance on an alliance with the navy. The fragility of the arrangement was revealed in December 1953, when the first sea lord, Admiral Sir Rhoderick McGrigor, secured the RAF's support for the Fleet Air Arm on the basis that both services would co-operate to effect a rapid run-down of the army. Fortunately for the army, the RAF's deal with the Royal Navy was to prove as brittle as its own. The government was not persuaded that aircraft carriers were less vulnerable than RAF ground bases, and McGrigor realized that he must build the senior service on securer foundations than air power. By 1955 it was once again backing the army on the issue of 'broken-backed' warfare. More importantly, it began to recognize that its best solution lay in securing a major nuclear role for itself. The major change in the 1952 strategy effected by Duncan Sandys's white paper of 1957—the proposal to switch the principal delivery system from the manned aircraft to the guided missile—proved to be the navy's opportunity. The strategy of 1957 was consistent with that of 1952, but for the RAF its implications were grievous.[28]

The most important consequence for the army of the 1957 white paper was the abandonment of conscription. Although during the early 1950s the chore of continuous training and its adverse effects on regular recruitment had turned many soldiers against national service, the army had done little to prepare itself for the change. Its political leverage had diminished both absolutely and relatively. 'The War Office never wins a battle', ran the inside view. 'The Admiralty has a strategy; the Air Ministry has tactics; the Army has neither.'[29] Whereas Mountbatten, the first sea lord, wooed Sandys with aristocratic charm and weekends at Broadlands, his army colleague, Sir Gerald Templer, could seem little more than petulant towards the defence minister. On one occasion, after a dinner party, he resorted to violence.[30]

Templer's views on the next war were consonant with Slim's. He believed that the communist threat would be met outside Europe, not within it, but that if there were war in Europe it would not be settled by a nuclear exchange. He queried the utility of nuclear weapons, and he continued to espouse the idea of a broken-backed conflict.[31] However, the most positive and politicized assertion of the army's position came not from Templer but from Lieutenant-General Sir John Cowley, the comptroller of munitions, in a provocative address to the Royal United Service Institution in November 1959. The 1957 white paper minimized the role of conventional forces, but Cowley seized on its lack of definition concerning lesser threats and Britain's continuing global commitment. He suggested that small forces could be used in the early stages of a crisis to prevent it growing, and that the army might itself be equipped with tactical nuclear weapons. He was using the ideas of graduated deterrence to ensure a continuing role for the army in national strategy. Furthermore he timed his attack for a moment of ministerial weakness. Sandys had gone, and his successor, Harold Watkinson, was still feeling his way. Watkinson responded by stressing that in future statements on defence policy from serving officers should have his approval, but Cowley's misdemeanour did not affect his career: he was elevated to the Army Council within a month.[32]

Sandys treated the chiefs of staff with a contempt that shocked them. The political clout of the individual services had waxed over the previous decade: the service ministers had seen it as their task to support their chiefs of staff, and the minister of defence had lacked the authority or the means to rein them in. But, confronted with a strong minister, the chiefs of staff responded by falling on each other rather than by rallying for mutual support. The travails of the RAF produced little sympathy in the army, Templer turning on the chief of the air staff, Sir Dermot Boyle, for advocating an independent British nuclear deterrent.[33] Although he enlisted Mountbatten's support in this attack, he did not give him his trust. On one occasion Templer told the first sea lord, 'Dickie, you're so crooked that if you swallowed a nail you'd shit a corkscrew.'[34]

Sandys's authority depended on two factors. The first was the backing of the prime minister, Harold Macmillan. As this ebbed, and as Conservative doubts mounted, the minister of defence's position weakened.[35] The second was his ability to divide and rule. The power of the individual services enabled Sandys to play off one against the other. Co-ordination of the type achieved by the chiefs of staff under Alan Brooke's chairmanship would have weakened his authority, not enhanced it. Ironically, neither he nor the chiefs of staff recognized this.

Sandys and Macmillan, like others before them, were driven by the need for economy and the recognition of the waste which overlapping and competing forces generated. They were also anxious to produce a strategy for national defence that was more than the product of the self-serving arguments of one particular arm. They feared that the advice given by the chiefs of staff reflected not the best collective opinion but the result of bureaucratic collusion—either the triumph of two services over the third, or the only workable compromise between all three.

In 1955 Macmillan's predecessor as prime minister, Anthony Eden, had appointed a chairman to the chiefs of staff committee who was over and above the service chiefs.[36] The government's hope that the chairman would provide it with a collective view was dashed by the hostility to the new arrangement among the chiefs of staff. The chairman, Marshal of the Royal Air Force Sir William Dickson, was supported by Mountbatten, if for no other reason than that the first sea lord was a committed centralizer. But Boyle and Templer became the spokesmen of the corporate virtues of the chiefs of staff. Templer stoked the fears of some Conservative leaders that in the hands of an extreme left-wing government the chairmanship could become a political appointment.[37]

The commitment of Macmillan and Sandys to centralization could only be confirmed by the obstructiveness with which their defence policies were confronted. Dickson became the first chief of the defence staff in 1958, and in 1959 he was succeeded by Mountbatten. Mountbatten was twice renewed in office, and his tenure of six years was crucial to the establishment of the new arrangements. By acquiring the joint planning staff, he ensured that the chief of the defence staff had some measure of operational control, and between 1960 and 1962 he created tri-service supreme commands in the Near East, Middle East, and Far East. In 1962 he proposed the creation of a ministry of defence organized on functional lines. All three service chiefs now turned against him.

Macmillan defused the situation by referring the scheme (on Mountbatten's advice) to two generals, Lord Ismay and Sir Ian Jacob. Both were veterans of the defence secretariat of the Second World War, and could be calculated to support change. In 1948 Jacob had sown the seeds of radical integration in Montgomery's mind and had written eloquently to the editor of the *Sunday Times* on the vices of 'a defence plan, which is merely the sum of those proposals from each service which others judge to be comparatively harmless to their own interests'.[38] But Jacob embodied more than the lessons of the Second World War. It may be not over-fanciful to see him as a symbol of Robertson's continuing influence, since the latter's son was one of his oldest and closest army friends. Jacob was the driving force behind the report which bore his and Ismay's names. Their solution, adopted in 1963, was to create a single ministry, which would bring the service departments under one roof and subordinate them to an overall control. But they rejected, at least for the time being, the full integration of the defence staff to which Mountbatten aspired. In this they seemed to shelve the more provocative of Jacob's instincts in order to ensure the acceptance of what they proposed. They reflected Slessor's point that the men who advised the government on the strategic employment of the armed forces should be those who had the final responsibility for the consequent operations conducted by those services.

The centralization of defence carried out between 1958 and 1963, and associated with Mountbatten's name, is doubly significant. First, its impulse was sustained by servicemen. The armed forces as much as the politicians, and certainly more than the civil servants, were responsible for the creation of the ministerial structure to which they were technically subordinate. The notion of fusion had developed to a point where the professional head of the services seemed to carry more weight than their political head. Certainly Macmillan's enthusiasm for centralization remained crucial, but Sandys's handling of defence matters created ripples within the Conservative party sufficient to ensure his removal in 1959. His successor, Harold Watkinson, only once rejected the advice of the chief of the defence staff in two and a half years, and when Watkinson went Macmillan asked Mountbatten himself whether he would like to be minister.[39] Mountbatten refused, but the combination of continuity in office and compliant politicians gave him an authority in domestic and international politics that none of his successors has been able to emulate.

The second significant feature was, however, the obverse of the first. Mountbatten's success in ministerial centralization was not reflected in the organization of the chiefs of staff. Each service maintained its own promotional hierarchy, and each service chief retained his right of appeal to the prime minister. Both the army and the Royal Air Force sustained their opposition to functional organization. Templer's successor but one as chief of the imperial general staff, Sir Richard Hull, had obstructed Mountbatten's efforts to create an integrated Far Eastern command. Once in Whitehall himself, Hull took up Templer's battle-cries, incorrectly citing Hitler's *Oberkommando der Wehrmachts*, which had clashed with the supreme command of the German army, as evidence of the dangers of centralization. In 1964 Montgomery wrote to Watkinson's successor as minister of defence, Peter Thorneycroft, to warn him of Hull's opposition to a unified ministry of defence. Hull was 'reactionary, single Service minded and has no wide progressive outlook or statesmanship'.[40] But in 1965 Hull succeeded Mountbatten as chief of the defence staff. The impetus for centralization stalled.

Denis Healey, who became minister for defence when the Labour party gained power in 1964, held that office for six years. He can claim to have been its most successful incumbent to date. Power shifted back from the professional to the politician. And it is reasonable to argue that it did so precisely because Mountbatten's schemes for the further integration of the defence staff were not implemented. As Mountbatten's tenure drew to a close, he renewed his efforts to create a genuine defence staff. Healey stalled. He did nothing to delay Mountbatten's departure or oppose Hull's succession. He acted in this way precisely because he felt that central control of defence policy should reside with the secretary of state and not with the chief of the defence staff.[41]

In effect Healey appreciated, as the Conservatives had not done, the opportunities which inter-service rivalry accorded the minister of defence to divide and rule. 'The competition for money', he wrote in his memoirs, 'required the senior officers of all three services to develop all the skills of the politician and the trade unionist. I sometimes felt that I had learned nothing about politics until I met the Chiefs of Staff. Each felt his prime duty was to protect the interests and traditions of his own service.'[42]

An indication of Healey's ministerial power was the approval of combined defence estimates in 1968 and their adoption in 1970–1. The consequence was the maintenance of balance between the three services. Over the long haul the army was the principal casualty. In 1947–8 the defence budget was unbalanced: the army commanded 43.4 per cent of the total, the navy 16.1 per cent and the RAF 15.5 per cent. But this was a reflection of the cost of conscription, of manpower's dominance over equipment. By 1964, with conscription gone, the army received only 26 per cent of the defence budget. The RAF share of the budget on the other hand, thanks to a succession of ultimately abortive projects including Skybolt and TSR-2, rose to 33.3 per cent in 1960–1. But over the shorter term, during Healey's tenure of the ministry of defence, it was the RAF that proved to be the main loser. By 1968 its share was running at 27 per cent for the third successive year, while the army's had recovered to 31 per cent.[43]

The army therefore did not feel threatened by Healey, for all the Labour government's decision to withdraw from east of Suez. Healey, himself a wartime soldier, paid frequent tribute to the professionalism and competence of the service. He sustained and promoted the careers of its most able senior officers, notably that of Michael Carver, who in 1973 became the army's second chief of the defence staff.[44] Above all, Healey made clear that Britain's prime defence obligation was to NATO, and by elevating the Continental commitment to the pole position in British strategy ensured the army's security in the services' pecking order for the foreseeable future.

Healey's battles were with the Royal Air Force and, most publicly, the Royal Navy. Mountbatten had argued the navy's case effectively. He had won Duncan Sandys over to the concept of the aircraft carrier, and he had secured the adoption of the nuclear-powered submarine. He had begun to recognize the potential of the submarine-launched ballistic missile, and the possibility that the British deterrent could be manned by the Royal Navy rather than by the Royal Air Force. His successors on the other hand became fixated with matters on the sea or above it, rather than with developments beneath it. They pushed for the modernization of the navy's aircraft carriers. The RAF argued that land-based aircraft could do the jobs hitherto done by the navy, and this controversy was in turn fed by the competing requirements of the navy and air force in aircraft development. Both the first sea lord, Sir David Luce, and the naval minister, Christopher Mayhew, resigned in 1966.

Mayhew's resignation was prompted principally by his belief that, if the navy did not get its carriers, Britain would not have the resources to sustain its commitments east of Suez. He was right—it was just that Healey cut the navy first and the commitments after, rather than vice versa. But Mayhew had a further point. He was protesting on behalf of service loyalties: centralization and the creation of a 'party line' stifled competition, depressed morale, and sapped initiative. His resignation was a symbol of the service minister's powerlessness, of his redundance once he became no more than the spokesman of his service, and of the waxing authority of the minister of defence. In 1967 the ministers of state for the individual services were downgraded to parliamentary under-secretaries.

The services' progressive loss of ministerial authority did not diminish the rivalries between them. This was because the staff structure was still divided. Each of the forces nominated the chief of the defence staff in rotation, and the suspicion remained that, however objective his chairmanship, he could not shed the loyalties formed in a lifetime devoted to a single service. Thus the debates over British strategy in the last decade and more of the Cold War continued to show the link between strategic thought and inter-service lobbying within the defence budget.

In 1980 the Conservative government announced its decision to replace the ageing Polaris submarine-launched missiles, which Britain had acquired from the United States in 1962, with the new American missile Trident. Like Polaris, Trident was a submarine-launched system, and would therefore be controlled by the Royal Navy. Public discussion before the announcement was extraordinarily restricted. The first really extensive and thorough press coverage was an article in *The Economist* for September 1979 by Admiral of the Fleet Lord Hill-Norton. Hill-Norton argued in favour of continuing and renewing Britain's deterrent. As a former chief of the defence staff, his case had the appearance of objectivity; as a former first sea lord, it carried a whiff of service self-interest. With the loss of carriers and the focus on Europe, the submarine-launched ballistic missile had become the means by which the navy might sustain an element of its former pre-eminence in British strategy.

When the government's announcement was made, the Americans had yet to decide whether to adopt the Trident I (or C4) missile, or Trident II (or D5). The latter had the greater range, the greater kilotonnage, and the larger number of warheads, but it was a system still under development and it would require a submarine twice the size of those in the Polaris fleet. The exact difference in costs was uncertain, but the D5 would clearly be more expensive, even if there were long-term savings in following whichever course the Americans chose for themselves. The chairman of the committee appointed in 1981 to review the choice was the first sea lord, Admiral Sir Henry Leach.[45] He could not be a dispassionate judge: he naturally wanted the biggest and the best for his service. Unsurprisingly, Britain followed the United States and adopted the D5. There remained the possibility that Britain might settle for fewer missiles per boat, and fewer warheads per missile than the Americans: rumour had it that on both counts the navy, and in particular Admiral Lord Fieldhouse, first sea lord from 1982 to 1985 and chief of the defence staff from 1985 to 1988, was ranged on the side of the maximum.

The debate about Trident—about the choice of a ballistic missile over a cruise missile, about the numbers of submarines and of warheads, about the need for a specifically British deterrent, about the independence of a system developed by another power—trickled on throughout the 1980s. Hill-Norton reiterated his commitment to a British nuclear deterrent.[46] But during the mid-1980s the cudgels were really taken up by another sailor, Lord Lewin, whose public reputation as chief of the defence staff was deservedly high after his performance in the Falklands war in 1982. At the end of the following year he declared that the cancellation of Trident 'would amount to unilateral disarmament'.[47] In February 1985 he wrote a major article in *The Times* supporting Britain's acquisition of Trident, its decision to go for the D5 missile, and rebutting the advocates of cruise missiles. He couched his arguments entirely in terms of the threat from the Soviet Union, but his advocacy extended beyond the end of the Cold War. By 1991–2 he presented its rationale as 'to insure against nuclear blackmail in an increasingly uncertain and unstable world', and to that end four, not three, boats remained essential.[48]

The Royal Air Force, although no longer the custodian of Britain's strategic deterrent, was still the possessor of sub-strategic nuclear systems, and tended, at least publicly, to support the navy.[49] The army, however, was at best neutral and in one case in particular deeply sceptical.

Field Marshal Lord Carver, who retired as chief of the defence staff in 1976, emerged as a strong opponent of Britain's possession of nuclear weapons. While in office Carver had been prepared to go along with naval and scientific pressure to update the Polaris warhead—the costly project for multiple re-entry vehicles codenamed Chevaline. Although uncertain of the value of an independent strategic force, he was prepared to accept the view that it was the navy's affair provided it had no impact on spending on conventional defence.[50] One of the difficulties for Carver and for other critics of Trident proved to be just this—the persistent but perverse claim that the nuclear deterrent carried no 'opportunity costs' for other elements of Britain's forces.

It would be stretching a point to say that Carver recognized the budgetary implications for conventional forces (and hence the army) first, and developed his strategic arguments thereafter. None the less the association between service self-interest and strategy existed as surely in his case as it did in that of the navy. The difference lay largely in the quality and intellectual rigour of the expression.

Carver did not reject the collective NATO need for nuclear deterrence. What he could not accept was the official position that London's acting as 'a second finger on the trigger' enhanced deterrence and cemented the alliance. The United States provided the nuclear guarantee for all and the idea that Britain might use nuclear weapons independently was for him incredible. British possession, rather than uniting Britain to America, implied distrust. Less theoretically and more seriously, Carver argued that a new British nuclear force would undermine NATO as a whole by drawing resources from other more important defence priorities. 'Maintaining a strong, mobile, armoured army, supported by a modern tactical air force, as our fair share of Nato's forces in the Central Europe sector, is a far more effective contribution both to a deterrent in war and to containing the enemy's forces if that fails than either duplicating the vast strategic nuclear strike capability of the United States or keeping afloat anti-submarine helicopter platforms.'[51]

Carver was not the only general to voice in public his opposition to Trident: in due course both Field Marshal Sir Nigel Bagnall, chief of the general staff between 1985 and 1988, and General Sir Hugh Beach did so too.[52] The army's strategic arguments for the improvement of conventional forces, particularly when in Carver's case they were accompanied by suggestions that the Royal Air Force might be abolished as a separate service, smacked of being *parti pris*. But they did of course conform, as surely as had contributions from other soldiers, to the ideas of graduated deterrence and flexible response. Furthermore in the early 1980s the operational doctrines developed from the so-called 'emerging technology' of precision-guided munitions suggested that in the future conventional warfare might prove sufficient both as deterrence and defence. Carver's position seemed logical and prescient.

Although the Conservative government of Mrs Thatcher could not accept Carver's case against a British nuclear deterrent, in other respects it endorsed the army's position. It inherited a defence budget whose constraints had propelled the chiefs of staff into political action. In December 1976, they had appealed to the prime minister—a move seen by two well-informed commentators as 'a lethal and hence successful threat' given the government's lack of an overall majority in the Commons.[53] The chiefs were successful in forcing Labour to hold defence spending steady for the time being, but they could not gainsay long-term economic realities. Fred Mulley, the Labour defence secretary, initiated a review, 'The Way Ahead'. The press was sufficiently aware of the tensions between the government and the chiefs of staff to suggest that the former was gagging the latter.[54] But, if that was the case, it did not stop the chiefs' discontent over service pay bubbling into the public arena. The Conservatives were both the beneficiaries and the victims of these tensions—beneficiaries in the sense that the services welcomed their promise of robustness in defence, victims because the chiefs' political manœuvres had thwarted Mulley's review, leaving his Conservative successor with the need to follow it through.

John Nott responded to the challenge. His own review, 'The Way Forward', was published in June 1981. The chiefs of staff had made a virtue of balance; Nott broke that principle, albeit not to the degree that the protests he elicited suggested. He decided that—after allowance had been made for the nuclear deterrent and home defence—priority should be given to the central European front ahead of the eastern Atlantic. The review's principal victim, therefore, was the Royal Navy. Chatham dockyard was to close, the number of frigates in commission was to be reduced from 59 to 50, and the number of aircraft carriers cut from three to two.

Nott's review hit equipment more than manpower: because of the length of the procurement cycle its long-term implications were therefore more far-reaching than earlier reductions. His strength, like that of Sandys and Healey, rested on his ability to divide and rule. Each of the armed forces was invited to list its needs, and was thus set in competition with the others.[55] Under him the civil service and the scientific defence community, already evident as a 'fourth arm' in the debate on Trident, gained in authority.

The problem for the navy was one both of timing and co-ordination. If it protested too soon, the public and the press would be insufficiently attuned to the issues. If it protested after the cuts were published, it would be too late. In January Lord Hill-Norton led the way in an exchange with Corelli Barnett in the correspondence columns of *The Times*: he rejected Barnett's call for a concentration on western Europe and made the case for Britain's contribution to 'deterrence outside the Nato area'.[56] In May Keith Speed, the parliamentary under-secretary for the navy, spoke in public against the impending cuts. An unlooked-for effect was to cause the chiefs of staff to shelve their plan for a collective visitation to the prime minister.[57] Speed himself promptly lost his job. He recognized full well the parallel with Mayhew,[58] and so to all intents and purposes did Mrs Thatcher. She completed what Healey had begun. Service ministers were removed entirely, and were replaced by two ministers of state, one for the armed forces and one for procurement.

It seemed that British defence policy had been set for the foreseeable future, and that the priority of the Continental commitment in peacetime would favour the army and the air force to the detriment of the navy. But then, in March 1982, the Argentinians seized South Georgia and prepared to invade the Falklands. Lord Lewin, the chief of the defence staff, was in New Zealand. Sir Henry Leach, the first sea lord, seized the opportunity which General Galtieri had given him. On 29 March he held a meeting to consider the creation of a possible task force. The government was in disarray, facing an international humiliation that could only rebound to its domestic discredit. None of the defence secretary, the chief of the air staff (who was acting as chief of the defence staff in Lewin's absence), or the chief of the general staff was enthusiastic about mounting operations at such range. Leach on the other hand was assertive and confident. Uninvited but attired in his full uniform, he presented the prime minister with what she saw to be the salvation of her political reputation and what he hoped would be the salvation of the Royal Navy.[59]

The dispatch of the Falkland task force was a tremendous gamble justified by its outcome. It did not endorse the Royal Air Force's claims for the projection of land-based air power at sea. It did endorse, in defiance of pre-Falklands thinking, the viability of amphibious operations. Thus its benefits for the Royal Navy were less resounding than Leach might have hoped. The Royal Marines, whom the navy might have been prepared to sacrifice in preference to more obviously maritime elements in the service, found their future assured.[60] The carrier programme was restored from two to three. But Nott himself was clear that the Falklands war did not constitute a case for 'an even wider maritime strategy'. He produced a ringing justification for the Continental commitment: 'unusually in our history we now have our forces already deployed in the right place', 'for the forward defence of Germany is the forward defence of Great Britain itself'.[61]

In making these assertions he was duly supported by a battery of generals, including Sir Hugh Beach, Sir Harry Tuzo, and Lord Carver. The last drew five lessons from the Falklands. Three were deeply inimical to the Royal Navy. 'The first is that an adequate force, land and air, on the ground is a much better assurance of defence than an operation to recapture what one has lost as a result of its absence . . . The second is what a long time it takes to move a force by sea . . . The third is the vulnerability of surface ships to air attack.'[62]

Nott left government in 1983. He had found his battles with the chiefs of staff, and particularly with the navy, bruising. While in office he had used a television interview to accuse the senior service of deliberately overbudgeting its forward expenditure.[63] Out of office he was even more forthright. He lambasted the navy for refusing to use the defence operational analysis unit because 'its scientific analysis and approach simply did not accord with naval prejudices'. More generally, he argued that the service staffs were 'excellent planners but . . . not normally conceptual thinkers'. His conclusion was pessimistic: 'Because of the disciplined and hierarchical nature of Service life, there is bound to be something of a negative influence on any radical forward thinking exerted by the most senior officers of each Service, who normally rise to the summit of their careers only after 30 years of service.'[64] Nott's desire was for a grand design to underpin British defence policy; his implication was that the service staffs were effectively undermining such ambitions.

The corollary of the 'functionalization' of the civilian side of the ministry of defence, completed by Mrs Thatcher's government in 1981, was a comparable reform of the military elements. Nott's declared aim had been to determine the balance between the forces 'in terms of real defence capabilities rather than on the outcome of a debilitating argument over each service's budgetary share'.[65] Instead the Royal Navy's obstructiveness had forced him back to the tactics of divide and rule. He may have used them to not inconsiderable effect but he was not enamoured of them.

What enabled Nott to effect a large measure of inter-service consolidation was a change of heart in the chiefs of staff. The awareness of their own vulnerability began finally to impinge. Sensitivity to Healey's exploitation of their divisions had prompted the navy and the air force to negotiate a concordat in 1970. During the ensuing decade the growth in power of the permanent under-secretary, and the establishment first of a procurement executive in 1971 and then of a financial management and planning group in 1977 meant that economic factors became more clearly articulated than strategic. The deputy chiefs of staff were removed in 1968 (and vice-chiefs followed in 1984) with the result that the uniformed element on the service boards diminished in relation to civilian representation.[66] In 1978 the chief of the defence staff, Marshal of the RAF Sir Neil (later Lord) Cameron, formed the 'Way Ahead Study Group' in anticipation of the next round of cuts. Its aim was to ensure an inter-service input in what became the 1981 review. Cameron's successor, Lewin, went one stage further. Dismayed by the fragmented response of the services to Nott's review, he argued that the chief of the defence staff should have his own central staff, and so be given the power to develop strategic concepts independently of the three armed forces. Lewin had served in the ministry of defence in the last stages of Mountbatten's stewardship, and was strengthened by the fact that the chief of the general staff, General Sir Edwin Bramall, had also been on Mountbatten's staff.[67] Through the Nott–Lewin reforms of 1982 the chief of the defence staff became less of a chairman representing the often discordant views of three independent services, and more the main strategic adviser of the minister of defence. He was given a deputy to co-ordinate the requirements of all three services, and field commanders became answerable to him rather than to the chiefs of staff committee. In the Falklands war, although the service chiefs had an input, only the chief of the defence staff attended meetings of the war cabinet. The armed forces had found a balance to the permanent under-secretary. They had also begun to find a collective voice.[68]

The process of unification was carried on by Nott's successor Michael Heseltine. In March 1984 he declared his intention 'to create a combined defence staff, responsible under the Chief of the Defence Staff and the Permanent Under-Secretary, for advising me on defence policy, military priorities and the conduct of military operations'.[69] The chiefs of each of the armed forces would lose any operational functions and would be removed from the centre of policy-making to its peripheries: their surviving role would be that of managing each of their services.

Heseltine planned to go further and faster than even the supporters of greater inter-service co-ordination were prepared to accommodate. Part of the problem was his style: he had formed his views in advance and then discussed them only with a small inner group. The chiefs of staff themselves were given four days, of which two were Saturday and Sunday, to consider the proposals before they were made public. Bramall, now chief of the defence staff, argued that in the formulation of policy (as opposed to the exercise of command) four minds were better than one, and that much of his own authority derived from the awareness that it had the backing of the three service chiefs.[70] His objections were echoed by three of his predecessors in office, including Lord Cameron. Cameron's main concern was morale: men should be commanded by members of the same service. His views, expressed in the press and parliament, were reinforced by Lords Hill-Norton and Carver, both arguing that policy and management should not be divorced. Only Lord Lewin spoke out forcefully in support of the changes.[71]

In the event the opposition of all three existing service chiefs, as well as that of the chief of the defence staff, was more muted than the public debate suggested. The Nott–Lewin reforms had broken the back of strong service opposition, and they had been tested by war and justified by victory. None of the chiefs could prepare an alternative model, and all recognized the advantages to their own service of a balanced budget and of a co-ordinated approach to strategy. Heseltine's main public critics—Lords Cameron, Hill-Norton, and Carver—had each in his own time supported greater inter-service staff co-operation. The door to fusion, even if it had taken forty years to inch open, was now more ajar than the single-service rhetoric suggested.[72]

The three service chiefs of staff prepared a joint memorandum expressing their reservations over Heseltine's plan, to which Bramall also appended his signature. At the beginning of July they exercised their right to see the prime minister. The concessions that they extracted were minimal. In his parliamentary statement Heseltine emphasized the importance of the service chiefs to 'the maintenance of the fighting effectiveness and the morale of their services', and he pointed out that they would have access to the defence staff as well as retaining 'substantial' staffs under their own control (in fact they were to be halved). For her part, the prime minister agreed to see the chiefs of staff on a more regular basis: thus an instrument which could be used to signal alarm by virtue of its exceptional nature became—from 1985—a matter of annual routine. On 19 July Bramall issued a statement to the effect that 'the chiefs of the Armed Forces are unanimous that a reorganization of the Ministry of Defence, announced yesterday by Mr Michael Heseltine, the Secretary of State for Defence, can be made to work extremely well'.[73]

What happened in 1984 seemed to be the triumph of civilian control over the military. The appointment of the chief of the defence staff was no longer the consequence of 'buggins' turn'—the process by which each service took the post in rotation, and filled it with its existing chief. Instead the secretary of state for defence made a recommendation to the prime minister, thus putting the final selection in the hands of the politicians, not of a single service. In 1991 Field Marshal Sir Richard Vincent, who had never been chief of the general staff, but had been vice-chief of the defence staff, got the job. Ironically, the image which Bramall conjured up was not so much one of Clodian military subordination, but its obverse—a fear that the chief would be appointed for political rather than professional reasons.[74] Neither extreme was really warranted: the fundamental trend lay in the middle—with the furtherance of fusion and integration.

The other half of the Heseltine reforms was the promotion of Thatcherite ideals of management throughout the ministry of defence. Commanding officers in the services became responsible for their own budgets. In some departments civilians became answerable to servicemen, just as in others servicemen were subordinate to civilians. Sir Ewen Broadbent, second permanent under-secretary of state in the ministry until 1984, wrote when reflecting on the changes:

The relationship between Ministers and the senior military staff underlines the need for the involvement of both groups in all parts of the Ministry. There is not a purely political or a purely military field. The managerial roles of the Service Departments still call for political guidance and an awareness of political sensitivities. Decisions of major international relationships must take account of military considerations and vice versa.[75]

In collective terms, therefore, all three services gained through the creation of a defence identity. A more homogeneous ministry allowed the military to acquire a measure of responsibility for its own administration as much as it permitted the introduction of the principles of 'new Conservatism'.

The Heseltine reforms brought financial advice out of the individual service departments and vested it in a powerful and centralized office of management and budget. The latter appeared to have the power to control future defence strategy by the selection of the financial information it decided to reveal.[76] Therefore, the key question for the individual services was whether in the next defence review centralized decision-making would force unequal reductions in the armed forces. The end of the Cold War and the collapse of the Soviet Union confronted them with this challenge sooner than they had anticipated. In the event their worst fears were not realized.

The first reason for this was that the reductions implicit in the phrase 'peace dividend' were less radical in intention and less immediate in their implementation than many expected. Tom King's 1990 review, 'Options for Change', was an interim position. Four years elapsed before Malcolm Rifkind followed it with 'Front-Line First'. Both secretaries of state were able to present their reviews as driven by the Treasury. They thus largely avoided the internal wrangles that had characterized both the Sandys white paper of 1957 and Nott's 'The Way Forward' of 1981. The service chiefs were by and large content to limit their objections to their usual meeting with the prime minister rather than to convene a special one.

A major reason for their comparative quiescence was that the service chiefs were marginal to both reviews. 'Options for Change' was driven by a small group notable for their collective rather than their single-service loyalties. The two servicemen were both soldiers, and the senior of them was Sir Richard Vincent, at that stage chief of the defence staff elect.[77] There was no attempt to include a sailor or an airman for the sake of balance. 'Front-Line First' was led by thirty-three inter-departmental study teams which sought ideas from throughout the services: an approach which threw up over 3,000 proposals from middle-ranking officers and civil servants. More than any other review it was driven from the bottom upwards as well as from the top downwards; more than any other review it was a symbol of fusion—military and civilian as well as inter-service—in its formulation.[78]

In the entire process no one strategic role was abandoned. In previous debates, each service had rationalized its claim on resources in the language of its own strategic functions. But in both the 1990 and 1994 reviews, and in the former in particular, great care went into the maintenance of the full range of defence capabilities. Not even in the 1980s, with the emphasis on western Europe, had the defence budget really departed from the principle of equal shares for all. Thus the main worry of the service chiefs about the creation of integrated defence structures was obviated.

'Options for Change', while halving the British army of the

Rhine to 25,000 men, still retained the Continental commitment. It cut the navy from 48 to 40 destroyers and frigates, and from 27 to 16 submarines, but it did not impugn the general principles on which the surface fleet rested. The Royal Air Force had a tougher ride, particularly under Rifkind. The tactical air-to-surface missile was shelved in 1993, its WE 177 free-fall nuclear bombs being abandoned without replacement, and its manpower was cut by 7,500 in 1994. If there was to be a backlash it was most likely to come from this, the service that had customarily used strategic arguments for political effect. But significantly the chief of the air staff attacked the Treasury, not the ministry of defence, and even concluded by apologizing to the chancellor of the exchequer.[79] His final position was to express gratitude that the cuts had not been worse. Moreover, nobody was disputing the importance of air power *per se*, and its more extreme advocates were not slow to deploy the evidence of the 1991 Gulf war in terms redolent of the strategic bombing theorists.

Part of the reason for service passivity was that the speed of the end of the Cold War had left the armed forces bereft of alternative ideas. 'Options for Change' was drawn up before NATO had decided its own strategy, and in referring to concepts tended to look backwards, not forwards. Then both the Gulf War and the Bosnian crisis intervened: global functions in support of the United Nations prompted a redefinition of what lay 'out of area' as far as NATO was concerned. For the army this revival of an extra-European role awakened a continuity that stretched back through counter-insurgency to colonialism. A more assertive strategy became possible with the end of the Cold War, and was publicly welcomed by a number of retired generals, including Sir Frank Kitson, Sir William Jackson, and—as he had now become—Field Marshal Lord Bramall.[80]

But such ideas were not the monopoly of the army. Amphibious operations, especially at extended ranges, clearly gave the navy an ability to lay to rest what one field marshal called 'the ghost of the battle of the Atlantic'.[81] It opened the door to an inter-service approach to strategy at a time when defence organization was itself forcing fusion. In 1994 a joint service command headquarters was established on a permanent basis, thus making the chief of the defence staff the effective commander-in-chief of all three services. At the same time the decision was taken to co-locate the command and staff training of all three services at Camberley. Although the latter decision certainly created unhappiness, what was more striking was the comparative lack of single-service protest.

As the twentieth century drew to its close, therefore, inter-service rivalry was muted, although certainly not extinct. Over the previous hundred years the services had expended so much of their political energy in battling with each other that they had often weakened themselves in their exchanges with their political masters. The creation of a collective defence identity might have betokened a reversion to open conflict between the services and the civilians in Whitehall. But this was not necessarily the case. The new machinery provided structures which meant that tri-service energies could be directed along approved channels, and which obscured the precise authorship of major decisions—whether civil or military.

The army no less than the other services ensured its political voice in the new arrangements. But it was equally clear that if it was to exercise the power thus vouchsafed it it had to establish a greater sense of cohesion than its own internal wrangles over the last century suggested it possessed. Its response to the 1990 and 1994 reviews was to look more to the future of its regimental structure than to its collective role. The need to redefine the Continental commitment, the re-emergence of alternative concepts—these were the issues which the army failed to address with sufficient urgency.

Conclusion

For all that this book has argued that the behaviour of the British army is as inherently political as that of other armies, there still remains the point that it has not staged a *coup d'état* nor been in the habit of making and unmaking governments. Many armies in other states have not been so restrained.

A simple, even simplistic, explanation for this difference is that the British army has rarely, if ever, had adequate motivation for massive political intervention. A professional army might reasonably feel disgruntled if it is held in low esteem by the society of which it is a part. It might also take exception if defence budgets are consistently criticized and if its professional wisdom on what is needed for the nation's security is therefore flouted. The British army has certainly felt aggrieved on both counts over the last hundred years, but a comparative approach will help put its sense of frustration in perspective. On the international scale the roots of its dissatisfaction are not significant.

By the 1890s the British army benefited from its association with the empire; it was then victorious in two world wars; and its conduct of colonial withdrawal was also deemed a success. Even in the early 1970s—with soldiers on the streets of Ulster and military units deployed at Heathrow, with student radicalism at its height and the military's image buffeted by America's performance in Vietnam—the army's legitimacy was never challenged. In an opinion poll conducted in 1972, 70 per cent of the respondents (of whom 60 per cent were aged between 16 and 24, and were therefore of student age) agreed that the armed forces were essential.[1] Since then the wars in the Falklands and the Gulf, both victorious and—equally important—brief, have enhanced its image. Between 1977 and 1982 only a quarter of respondents to opinion polls thought that too much was being spent on defence.[2] In 1995 the foreign secretary, Douglas Hurd, in an effort to relaunch Britain's status as a major player in international relations, trumpeted the quality and professionalism of its armed forces as a principal component of that bid.

In general terms, therefore, the army has been held in high regard, its constitutional status—despite the legacy of the Mutiny Act—unimpugned. There has been no parallel with the German army's alienation from its parent society in 1918. Seeking to divest itself of responsibility for defeat, it claimed that it had been stabbed in the back. The British army escaped too the fate of France's army in May 1940. Its collapse produced an internal division which outlived Vichy France and spawned violent differences in the withdrawal from empire.

Furthermore, over the same period, Britain has been a consistently high spender in terms of defence. Before 1914 it allocated more per head of the population to defence than did either France or Germany.[3] In both world wars it spurned financial orthodoxy in its bid to create, among other things, a mass army: it effectively bankrupted itself rather than stint on war expenditure. Within NATO Britain has consistently spent a greater percentage of its gross national product on defence than has any of its European allies. Of course there have been major battles about the army estimates, and more than in other powers the army has had to compete with the navy for its slice of the total, but it is none the less important not to confuse the rhetoric of those battles with the reality. Certainly more might have been spent, but equally certainly a great deal was spent.

What all this suggests is that to some extent Clode's expectations have been fulfilled. Rather than the civilians colonizing the military, the military have colonized the civilians. The army has enjoyed a far higher profile in twentieth-century Britain than the liberal—or perhaps it should be Whig—orthodoxy allows. Militarism is a word now loaded with too many meanings and productive of misunderstanding, but it is the only abstract noun that will do the job. If it is interpreted as a veneration of military values and appearances in excess of what is strictly necessary for effective defence, then it is not as inapplicable to Britain as the orthodoxy allows.[4]

That militarism has expressed itself not just in the army's popularity but also in Britain's fiscal management and in the relationship between it and the procurement of new technology. In the Napoleonic wars Britain preferred to adopt income tax rather than impose conscription. Under the premiership of Sir Robert Peel, it extended this preference into peacetime. The absence of compulsory military service for much of its history has been seen as a symptom of the country's liberalism and a cause of the army's low profile. But there has been a trade-off: although defence has not sequestered Britain's manpower, it has taken its disposable income instead. By 1914 Britain was the only country in the world with an effective system of income tax. At that date the principal purpose of central government expenditure remained the maintenance of the army and the navy. With the wealth thus vouchsafed it, the British government put itself in the forefront of weapons research and production, and it fought to sustain that primacy even when its wealth—at least in comparative terms—ebbed. In 1905 Britain led a revolution in warship design with the introduction of the dreadnought; it pioneered the tank; it established the first independent air force in the world, and in the 1930s it put its faith in the bomber. Since 1945 its attention to, and reliance on, nuclear weapons has been greater than any other medium rank power with the possible exception of France.[5]

State intervention has flowed more certainly from the demands of war industry and has been sustained more effectively by the research and development needs of weapons technology than it has been by the ideas of socialism. Furthermore, the establishment of a cosy, monopolistic relationship between arms suppliers and their end users goes back at least to the days of Jackie Fisher. The arms business may with justice feel that the state has been a capricious and uncertain customer, forcing it into export markets to survive, but its complaints must be balanced with its high profile in the recent history of British industry.

Moreover, during the course of the last century, and particularly since 1945, the army's subordination to parliament has become a constitutional figment rather than a practising reality. The role of the House of Commons has been largely passive—to be informed of policy decisions once taken rather than to have an active say in the shaping and making of those decisions. 'There is no area of governmental activity', David Owen told the House of Commons in 1972, 'where parliamentary control is in reality less effective than in the area of defence.'[6]

The creation of a parliamentary subcommittee on defence and external affairs in 1971 went some way to meeting Owen's complaint. Its reports are perhaps the most valuable source of evidence on defence matters in the public domain. But the committee does not attempt to review the full range of the defence estimates, and the civil servants of the ministry have—on their own admission[7]—become proficient in stonewalling lines of inquiry and in producing obfuscatory answers to straight questions. When, in December 1976, the chiefs of staff exercised their right of direct access to the prime minister, the content of their complaints remained unknown to either the House of Commons as a whole or to the select committee in particular. As the Labour MP and constitutional expert Professor John Mackintosh pointed out, the House was being asked to vote on the defence estimates without any knowledge of the strategic arguments that so worried the chiefs of staff. 'In every other major legislature in the western world', he said, 'there is a defence committee which would have summoned the Chiefs of Staff, heard their objections and cross-examined them, so that the M.P.s would know what were the issues involved.'[8]

It seems that the political controls on the army have loosened. And yet it has been a principal argument of this book that what has restrained the army from political intervention has been the strength of the governmental framework within which it has had to operate. The general background to the lack of military–political friction in twentieth-century Britain has been that the nation and the state have gone a long way towards accommodating the needs of the services in particular and of defence in general. The mechanism that has made for this degree of co-operation has been the establishment of integrated control. The push for fusion, culminating in the creation of the Ministry of Defence, has blended civilian with military. The ministry, instead of seeing itself as the check on the services as did the eighteenth-century War Office, has emphasized the collaborative nature of the enterprise. Cuts are imposed by the Treasury against the wishes of the civilian administrators within the ministry. Those civilians themselves identify increasingly with the services. Ministers see nothing incongruous, let alone constitutionally unsettling, in donning flak jackets and combat fatigues: George Younger, as secretary of state, on occasion even wore a uniform with a specially designed ministry of defence badge for his beret.

Co-ordination and integration have made professional expertise more powerful. The tendency in most defence matters is to trust the expert.[9] The occasions when the growth of integration has led to a fear of loss of civilian control have been remarkable for their rarity. In 1951, when Churchill returned as prime minister, his first instinct was to make himself minister of defence. Thwarted, his second was to militarize the administration of the services. Field Marshal Lord Alexander was appointed minister of defence, and Brigadier Anthony Head secretary of war. The leader of the opposition, Clement Attlee, criticized Alexander's appointment in the House of Commons: he drew the parallel with Kitchener, and emphasized the difficulties for a professional in having to control other professionals. But Attlee's complaints were somewhat disingenuous: in 1948 he himself had asked a sailor, Mountbatten, to be minister of defence.

The enhancement of the political position of the services through the processes of integration has been so unquestioned largely because Britain's strategic position has changed. The nostrums of liberalism and constitutionalism rested on the relative security afforded by the Channel. The two world wars came sufficiently close together to reinforce the point that Britain's defence now rested on the further side of the Channel, in Europe. The Cold War confirmed that conclusion. The effect was to keep Britain in a state of semi-mobilization, if not actual war, for much more of the twentieth century than was the case in the preceding hundred years.

Secondly, for most of that time it was a member of a military alliance. In both world wars it needed to speak with a single voice on the matter of strategy. Divisions between soldiers and politicians were likely to be exploited by Britain's allies for strategic or diplomatic advantage. Since the 1950s NATO has exerted a comparable pressure, and in particular has enhanced the power of the chief of the defence staff, who has become the government's professional spokesman in the alliance on matters of strategy.

New technology has of course been a key element in changing Britain's strategic position. The bomber and the missile have undermined the defensive strength of the Channel. Both weapons systems have proved symptomatic of further pressures for integration. The question posed by technology of a totally new order is that of fitting it into the existing patterns of service organization. Which of the armed forces should man it and be responsible for its development? And how should it be paid for? Co-ordination has been a way of seeking economies and avoiding the waste in duplication of effort.

As the twentieth century closes, Britain seems to have completed the protracted process of creating an integrated defence administration. It now has a defence staff that is more powerful than the individual services. A situation has been reached which sets a 'purple'-minded officer against the particularisms of his own navy blue, scarlet, or sky-blue service. The effect in public terms has been to defuse the defence debate—to concentrate professional concerns along channels internal to the ministry.

Such a situation gives rise to fresh dangers. One is of course the issue of public accountability. When in the past soldiers broke silence and voiced their discontents to the press, they informed the electorate of defence choices that otherwise remained obscure. They may have undermined the norm of political subordination, but they affirmed the potential of open government.

Another risk is that of the army's political emasculation. The belief, however mistaken, that the British army has been apolitical throughout its history may be used to suppress the political instincts to which its professionalism would otherwise give rise. Countless mess conversations bear testimony to this tension. One former regular officer, Sir John Baynes, wrote in 1991, 'The greatest danger is not that soldiers should start taking too much interest in what the politicians say or think about them, as the reverse'. His worry was that the soldier's contempt for the politician, his tendency 'to describe politicians as rogues or liars', would make the army politically ignorant.[10] This, after all, was once the interpretation—now rejected—used to explain the German army's subordination to the

Nazi party: in placing itself above politics in the 1920s, it had allegedly made itself politically naïve.[11]

In much of the received wisdom concerning the British army's place in politics, such an outcome could be seen as a laudable objective, not a possible jeopardy to national security. Should we therefore conclude that the army has at last become politically neutral—that the norm has become a reality?

The answer must be a resounding no. The potential for political activity on the part of the army remains as great today as it was in the past. That this is so is not because of some malevolent intent but—at the risk of excessive repetition—because it is in the nature of a professional army.

Lieutenant-General Sir Michael Rose was until 1995 the commander of the United Nations forces in Bosnia. Here is a classic example, familiar from the empire, of a soldier exercising authority with implications that are as much political as military. Furthermore, Rose had to do more than manage the situation within Bosnia; he had also to unite the sometimes competing roles of being a British officer, a member of NATO, and a United Nations commander. It was in this last capacity that he wrote to *The Times* on 2 November 1994, despite being a serving officer in the army and despite injunctions against those in such positions communicating with the press. His purpose in doing so was to emphasize that his task was that of peacekeeping, not of peace enforcement. In other words, he was using a public forum to declare his reluctance to employ NATO airstrikes against Bosnian Serbs, so infuriating not only Bosnian Muslims but also the government of the United States.

The most recent British commander in battle was General Sir Peter de la Billière. His book on the Gulf War, *Storm Command*, sniped at politicians in a fashion that is characteristic of the army's self-deception in regard to its own alleged political neutrality. In his autobiography, published in 1994, he went one stage further, declaring that 'politicians must have the courage to listen particularly to the advice of their military commanders'.[12] His grouse was given substance in January 1992 by the leak of a confidential military report criticizing the defence operations executive, the operational interface between the ministry of defence and the war cabinet. 'The lack of a clear, positive military lead from the D.O.E.', the report

said, 'led to excessive ministerial influence in military decisions.'
The ministers were more concerned with costs than with strategy.[13]
The sources of military–political friction in wartime have not
disappeared.

While the Gulf crisis unfolded, the army was nervously looking
over its shoulder at the implications for its future of the end of the
Cold War. In 1991, General Sir John Chapple, the chief of the
general staff, exercised his right to see the prime minister in order
to register his concern about the 1991 defence cuts—a fact duly
reported in the press. A letter which Chapple wrote to Tom King,
the minister of defence, was leaked. Three additional points are
worth noting about Chapple's protest, all of them suggesting elements of continuity in the army's management of politics. First, the
focus of Chapple's dissent was the regiment. Secondly, it was simultaneously made known in the press that the queen was unhappy
about the proposed amalgamations. Thirdly, he was backed up by
public protests from retired chiefs of the general staff and of the
defence staff.

In Chapple's case, the matters which preoccupied him were not
of great concern to an electorate anxious to reap what it imagined
was a 'peace dividend'. But the disposition to intervene so that
professional matters become the stuff of mainstream politics
persists.

In 1987 the chiefs of staff told the government that they would not
become involved in any election debate over the Labour party's
policy of unilateral nuclear disarmament. But at the same time the
press was able to publish the advice that the chiefs of staff would
give an incoming Labour government—to the effect that they
would exercise their right to see the prime minister and express
their reservations concerning a non-nuclear defence strategy. As if
this had not done enough to discredit any notions of political
subordination or of neutrality on the matter, the alleged reticence
of the services was further undermined by the public statements of
former chiefs of staff. Lord Lewin, predictably given his commitment to Trident, said he would have resigned as chief of the defence
staff rather than implement a non-nuclear strategy. His successor in
that office, Lord Bramall, was more cautious but still said that the
immediate removal of Polaris would weaken Britain's defences.[14]

When Mrs Thatcher's position as leader of the Conservative party was challenged during the Gulf war it was the same duo who spoke out. Lord Lewin, with naval directness, said that any change in leadership would be 'crazy'. Bramall was more circumspect but made a similar observation.[15] Perhaps the significant point to make is that, although the services were thus drawn in to a domestic political debate, their views had no appreciable impact on the outcome. John Major duly became prime minister.

The opinions expressed over Labour's defence policy or over the Conservative party leadership do not mean that the army has become committed to partisan loyalties—that it has fulfilled the expectation of the left by giving its support to the Conservatives. Its politics follow its professional priorities. In the autumn of 1991 General Sir John Akehurst, who had just retired as deputy supreme allied commander in Europe, wrote to the editor of the *Journal of the Royal United Services Institute for Defence Studies*: 'It is now clear that the Tories have handled the so-called "Options for Change" (what options?) with a lack of strategic clarity, manifest insensitivity and unfairness, and obvious political expediency, to say nothing of Treasury domination, and have thus put the country's security in jeopardy. The electorate will doubtless take note of this.'

The spirit of Henry Wilson, or perhaps in this case of Garnet Wolseley, lives on.

NOTES

Notes to Chapter 1

1. Carl von Clausewitz, *On War*, ed. and trans. Michael Howard and Peter Paret (Princeton, 1976), 87.
2. Ibid. 606.
3. Ibid. 607.
4. Ibid. 608.
5. Ibid. 608 n.1.
6. Carl von Clausewitz, *On War*, trans. J. J. Graham, new and rev. edn. by F. N. Maude (3 vols., London, 1911), iii. 127.
7. Peter Paret, 'Clausewitz: A Bibliographical Survey', *World Politics*, 17 (1965), 272–85; for the discussion that follows, see especially Christopher Bassford, *Clausewitz in English: The Reception of Clausewitz in Britain and America 1815–1945* (New York, 1994), 79–81.
8. G. F. R. Henderson, *The Science of War: A Collection of Essays and Lectures 1891–1903* (London, 1919), 12, 18–19.
9. G. F. R. Henderson, *Stonewall Jackson and the American Civil War* (2 vols., London, 1906), i. 231.
10. Ibid. 208, 215, 406.
11. Ibid., pp. xxi, 406–7.
12. Ibid. 406.
13. Ibid. 207–8.
14. Ibid. 215.
15. J. W. Fortescue, *Military History: Lectures Delivered at Trinity College, Cambridge* (Cambridge, 1923; 1st edn. 1914), 62, 98.
16. Ian F. W. Beckett (ed.), *The Army and the Curragh Incident, 1914* (London, 1986), 294.
17. Ibid. 295.
18. Ibid. 369–70.
19. Ibid. 364.
20. Michael Howard, 'The Armed Forces as a Political Institution', in M. Howard (ed.), *Soldiers and Governments: Nine Studies in Civil–Military Relations* (London, 1957), 21.
21. Correlli Barnett, *Britain and her Army 1509–1970: A Military, Political and Social Survey* (London, 1970), 130.
22. David Chandler and Ian Beckett (eds.), *The Oxford Illustrated History of the British Army* (Oxford, 1994), p. xvi.
23. Martin Edmonds, *Armed Forces and Society* (Leicester, 1988), 73.
24. Samuel P. Huntington, *The Soldier and the State: The Theory and Practice of Civil–Military Relations* (Cambridge, Mass., 1957), 97.

25. Ibid. 36.

26. Ibid. 47.

27. Morris Janowitz, *The Professional Soldier: A Social and Political Portrait* (1971 edn.), quoted in Lawrence Freedman (ed.), *War* (Oxford, 1994), 126.

28. Gwyn Harries-Jenkins, 'The British Armed Forces', in Morris Janowitz and Stephen D. Westbrook (eds.), *The Political Education of Soldiers* (Beverly Hills, Calif., 1983), 84.

29. M. Janowitz, *The Professional Soldier* (New York, 1960), 263; see also C. B. Otley, 'Militarism and the Social Affiliations of the British Army Élite', in Jacques van Doorn (ed.), *Armed Forces and Society: Sociological Essays* (The Hague, 1968), 86–7.

30. Huntington, *The Soldier and the State*, 33.

31. Morris Janowitz, 'Armed Forces and Society: A World Perspective', in van Doorn (ed.), *Armed Forces and Society*, 25.

32. Gwyn Harries-Jenkins, *The Army in Victorian Society* (London, 1977), esp. 217–75.

33. The literature on these points is too complex for detailed analysis here, but Karl Demeter, *The German Officer-Corps in Society and State 1650–1945* (London, 1965) is the starting-point. Subsequent scholarship has emphasized the inclusion of non-nobles within the officer corps, especially after 1912, and of course many aristocratic officers chose *not* to oppose Hitler in 1944.

34. See, for example, David Bien, 'The Army in the French Enlightenment: Reform, Reaction and Revolution', *Past and Present*, 85 (Nov. 1979), 68–98; Geoffrey Best, *War and Society in Revolutionary Europe, 1770–1870* (London, 1982), 23–7; Hew Strachan, *European Armies and the Conduct of War* (London, 1983), 65; Christopher Storrs and H. M. Scott, 'The Military Revolutions and the European Nobility c.1600–1800', *War in History*, 3 (1996), 1–41.

35. Gaetano Mosca, *The Ruling Class* (New York, 1939; 1st pub. in Italy in 1896). Mosca's only reference to the British army, which prefigures Janowitz, is to be found on p. 233.

36. Otley, 'Militarism and the Social Affiliations of the British Army Élite', 103.

37. Ibid. 106.

38. R. G. L. von Zugbach, *Power and Prestige in the British Army* (Aldershot, 1988), 3, 79–81, 86, 95–7.

39. Amos Perlmutter, *The Military and Politics in Modern Times: On Professionals, Praetorians, and Revolutionary Soldiers* (New Haven, 1977), 8.

40. Ibid. 2–3.

41. Ibid. 39–40.

42. Ibid. 8.

Notes to Chapter 2

1. Hew Strachan, *Wellington's Legacy: The Reform of the British Army 1830–54* (Manchester, 1984), 116–17; Michael Glover, 'The Purchase of Commis-

sions: A Reappraisal', *Journal of the Society for Army Historical Research*, 58 (1980), 223–35.

2. Edward M. Spiers, *The Late Victorian Army 1868–1902* (Manchester, 1992), 93–9; Edward M. Spiers, *The Army and Society 1815–1914* (London, 1980), 7–8.

3. Strachan, *Wellington's Legacy*, 110–11.

4. André Corvisier, *Armies and Societies in Europe 1494–1789* (Bloomington, Ind., 1979; 1st pub. 1976), 115.

5. John Childs, *The Army, James II, and the Glorious Revolution* (Manchester, 1980), 1–3, 27–8, 39–41, 45, 49, 71–2, 84–5, 156–62.

6. John Childs, *The British Army of William III 1689–1702* (Manchester, 1987), 7–8, 13–14, 70–7, 197–8.

7. John Brewer, *The Sinews of Power: War, Money and the English State 1688–1783* (London, 1989); for related points, see also Linda Colley, *Britons: Forging the Nation 1707–1837* (New Haven, 1992).

8. On Scots' careerism, see Colley, *Britons*, 126–31; James Hayes, 'Scottish Officers in the British Army, 1714–63', *Scottish Historical Review*, 37 (1958), 23–33; Keith M. Brown, 'From Scottish Lords to British Officers', in Norman Macdougall (ed.), *Scotland and War AD 79–1918* (Edinburgh, 1991).

9. Alan Guy, *Oeconomy and Discipline: Officership and Administration in the British Army 1714–63* (Manchester, 1985), 9, 19–20, 89–101, 137–41.

10. J. A. Houlding, *Fit for Service: The Training of the British Army, 1715–1795* (Oxford, 1981), 105, 109–15.

11. This has been a recurrent theme of my own work. See, in particular, *Wellington's Legacy*; *From Waterloo to Balaclava: Tactics, Technology and the British Army 1815–54* (Cambridge, 1985); 'The British Way in Warfare', in Chandler and Beckett, *Oxford Illustrated History of the British Army*.

12. Harries-Jenkins, *Army in Victorian Society*, 44.

13. Spiers, *Late Victorian Army*, 94.

14. Childs, *The Army, James II, and the Glorious Revolution*, 107–9.

15. Basil Duke Henning, *The House of Commons 1660–1690* (London, 1983), i. 10; Romney Sedgwick, *The House of Commons 1715–1754* (London, 1970), i. 155; Lewis Namier and John Brooke, *The House of Commons 1754–1790* (London, 1960), i. 138; R. G. Thorne, *The House of Commons 1790–1820* (London, 1986), i. 306–10.

16. Namier and Brooke, *House of Commons 1754–1790*, i. 138.

17. Thorne, *House of Commons 1790–1820*, i. 309.

18. Namier and Brooke, *House of Commons 1754–1790*, i. 141–2.

19. Thorne, *House of Commons 1790–1820*, i. 306–10.

20. Sedgwick, *House of Commons 1715–1754*, i. 141–2; Namier and Brooke, *House of Commons 1754–1790*, i. 139–40; L. B. Namier, *The Structure of Politics at the Accession of George III* (London, 1929), i. 31, 34, 37.

21. L. G. Johnson, *General T. Perronet Thompson 1783–1869: His Military, Literary and Political Campaigns* (London, 1957).

22. See Edward M. Spiers, *Radical General: Sir George de Lacy Evans 1787–1870* (Manchester, 1983).

23. John Vincent, *Formation of the British Liberal Party 1857–1868* (Harmondsworth, 1972; 1st pub. 1966), 42.

24. Harries-Jenkins, *Army in Victorian Society*, 220–30.

25. Charles M. Clode, *The Military Forces of the Crown: Their Administration and Government* (2 vols., London, 1869), i. 194.

26. Ibid. 161; also 192–4.

27. Spencer Childers, *The Life and Correspondence of the Right Hon. Hugh C. E. Childers 1827–1896* (2 vols., London, 1901), i. 282–3; Hansard, 3rd series, cclvii, cols. 324–5, 10 Jan. 1881; also cols. 1487–8, 27 Jan. 1881. Harries-Jenkins, *Army in Victorian Society*, 258, seems to me to misinterpret this episode.

28. Thorne, *House of Commons 1790–1820*, i. 313.

29. Ian F. W. Beckett, *Riflemen Form: A Study of the Rifle Volunteer Movement 1859–1908* (Aldershot, 1982), 153; Ian Beckett, *The Amateur Military Tradition 1558–1945* (Manchester, 1991), 192.

30. Nevil Macready, *Annals of an Active Life* (London, [1924]), 260.

31. W. L. Guttsman, *The British Political Élite* (London, 1963), 291.

32. Janowitz, *Professional Soldier*, 389.

33. Hubert Gough, *Soldiering on* (London, 1954), 204–5.

34. For illustrations of these points, see Robert Wright, *Dowding and the Battle of Britain* (London, 1969), Corgi edn., 202–3; Harold Nicolson, *Diaries and Letters 1939–1945* (London, 1967), 209.

35. Janowitz, *Professional Soldier*, 389.

36. Philip Abrams, 'Democracy, Technology and the Retired British Officer', in Samuel P. Huntington (ed.), *Changing Patterns of Military Politics* (New York, 1962), 166, 170.

37. Humphrey Crum Ewing, 'A Question for Ministers', *Strategic and Combat Studies Institute Occasional Papers*, 11 (1995), 18.

38. Brewer, *Sinews of Power*, 44.

39. Harries-Jenkins, *Army in Victorian Society*, 218; David Cannadine, *The Decline and Fall of the British Aristocracy* (New Haven, 1990), 520.

40. Guttsman, *British Political Élite*, 122.

41. The most revealing account remains T. F. Gallagher, 'Cardwellian Mysteries: The Fate of the British Army Regulation Bill, 1871', *Historical Journal*, 18 (1975), 327–48. See also Spiers, *Late Victorian Army*, 11–19.

42. Willoughby Verner, *The Military Life of H.R.H. George, Duke of Cambridge* (2 vols., London, 1905), i. 1–26.

43. Rhodri Williams, *Defending the Empire: The Conservative Party and British Defence Policy 1899–1915* (New Haven, 1991), 14, 46, 100–18.

44. Gregory Phillips, *The Diehards: Aristocratic Society and Politics in Edwardian England* (Cambridge, Mass., 1979), 82–110; Cannadine, *Decline and Fall of the British Aristocracy*, 48–53, 519–29.

Notes to Chapter 3

1. C. M.Clode, *The Military Forces of the Crown* (London, 1869), i. pp. iii–v, 84
2. Hansard, 1st series, xxxii, col. 986, 28 Feb. 1816.
3. T. B. Macaulay, *The History of England from the Accession of James the Second* (Popular edn., 5 vols. in 2, London, 1889), i. 678. For what follows, see also C. H. Firth, *A Commentary on Macaulay's History of England* (London, 1938), esp. 146.
4. Macaulay, *History of England*, i. 676.
5. Ibid. ii. 634.
6. Lois G. Schwoerer, *'No Standing Armies!' The Antiarmy Ideology in Seventeenth-Century England* (Baltimore, 1974), 51.
7. Ibid. 71.
8. On the militia, see J. R. Western, *The English Militia in the Eighteenth Century. The Story of a Political Issue 1660–1802* (London, 1965); Beckett, *Amateur Military Tradition*, 49–59, 62, 122, 152, 185–7.
9. Schwoerer, *'No Standing Armies!'*, 194.
10. Ibid. 151; Guy, *Oeconomy and Discipline*, 5.
11. Guy, *Oeconomy and Discipline*, 5–6; Peter Rowe, 'The British Soldier and the Law', in Martin Edmonds (ed.), *The Defence Equation: British Military Systems, Policy, Planning and Performance* (London, 1986), 175–6.
12. Fortescue, *Military History*, 60.
13. Childs, *British Army of William III*, 84–7.
14. Clode, *Military Forces of the Crown*, i. 261–2.
15. Childs, *British Army of William III*, 191–5.
16. This discussion is informed by, even if its interpretation differs from, John Robertson, *The Scottish Enlightenment and the Militia Issue* (Edinburgh, 1985), esp. 31, 69–70, 214–15.
17. Clode, *Military Forces of the Crown*, ii. 39; Hampden Gordon, *The War Office* (London, 1935), 39–40; 23 Geo. III, c. 50.
18. R. E. Scouller, *The Armies of Queen Anne* (Oxford, 1966), 5, 10–20; also Guy, *Oeconomy and Discipline*, 25–6.
19. Kenneth Bourne, *Palmerston: The Early Years 1784–1841* (London, 1982), 161.
20. Palmerston to Althorp, Feb. 1833, quoted ibid., 180.
21. Ibid. 161–80 contains a full discussion of these issues; Clode, *Military Forces of the Crown*, ii. 722–3 gives the text of the warrant; see also ii. 265, 338.
22. Bourne, *Palmerston*, 180.
23. Wellington to Melbourne, 25 Mar. 1837, Dalhousie muniments, Scottish Record Office, GD 45/8/66/6.
24. Clode, *Military Forces of the Crown*, ii. 242.
25. Namier and Brooke, *House of Commons 1754–1790*, i. 140–1.
26. Richard Evans Morse, 'Money or Merit? The Early Development of a Modern Officer Promotion Process in the British Army 1815–1830' (University of London Ph.D. thesis, 1977), 131; Bourne, *Palmerston*, 171.
27. Neville Thompson, *Wellington after Waterloo* (London, 1986), 67; Thompson is the best recent guide to what follows.

28. Ibid. 73, 80, 92–3; Norman Gash, 'The Duke of Wellington and the Prime Ministership 1824–30', in N. Gash (ed.), *Wellington: Studies in the Military and Political Career of the First Duke of Wellington* (Manchester, 1990), 117–18.

29. Canning to Wellington, 5 May 1827, cited by Gash, *Wellington*, 121.

30. Quoted by Bourne, *Palmerston*, 179.

31. Gash, *Wellington*, 179–80; also Thompson, *Wellington after Waterloo*, 200–1, 205, 223–8.

32. Palmerston to Althorp, Feb. 1833, quoted by Bourne, *Palmerston*, 180.

33. Strachan, *Wellington's Legacy*, 250–9.

34. Theodore Martin, *The Life of His Royal Highness the Prince Consort* (2 vols., London, 1875–6), ii. 253–62.

35. Strachan, *Wellington's Legacy*, 35–8.

36. Clode, *Military Forces of the Crown*, i. 97.

37. W. S. Hamer, *The British Army: Civil–Military Relations 1885–1905* (Oxford, 1970), 135–40.

38. John Vincent (ed.). *A Selection from the Diaries of Edward Henry Stanley, 15th Earl of Derby (1826–93), between September 1869 and March 1878* (London, 1994), 177.

39. Ibid.

40. Brian Robson (ed.), *Lord Roberts in India: The Military Papers of Field Marshal Lord Roberts 1876–1893* (Stroud, 1993), 400; Noble Frankland, *Witness of a Century: The Life and Times of Prince Arthur, Duke of Connaught 1850–1942* (London, 1993), 156, 183–4, 190–200.

41. Julian Symons, *Buller's Campaign* (London, 1963), 41.

42. G. H. L. Le May, *The Victorian Constitution: Conventions, Usages and Contingencies* (London, 1979), 81.

43. Frank Hardie, *The Political Influence of the British Monarchy 1868–1952* (London, 1970), 94–8.

44. Maurice V. Brett (ed.), *Journals and Letters of Reginald Viscount Esher* (4 vols., London, 1934), i. 358–9. See also Peter Fraser, *Lord Esher* (London, 1973), 87–91.

45. Brett, *Esher*, i. 369.

46. Ibid. 364–5.

47. Ibid. 408; see also Fraser, *Esher*, 18, 73, 89–91; Frankland, *Witness of a Century*, 227.

48. Hamer, *British Army*, 223–32.

49. David Dilks, *Curzon in India* (2 vols., London, 1970), ii. 103.

50. Hamer, *British Army*, 244, 252–3; Fraser, *Esher*, 101; Frankland, *Witness of a Century*, 231–9.

51. Brett, *Esher*, ii. 104, 163.

52. Beckett, *The Army and the Curragh Incident*, 140, 147–9, 326–31.

53. Brett, *Esher*, i. 391; see also ii. 50–1, 157.

54. Robert Blake (ed.), *The Private Papers of Douglas Haig 1914–1919* (London, 1952), 17, 19, 20, 36, 97–8, 108–11, 138. See also Peter Bryant, 'The Recall of Sir John French', *Stand to!* (1988), 22: 24–9; 23: 32–8; 24: 22–6.

55. David R. Woodward (ed.), *The Military Correspondence of Field Marshal Sir*

William Robertson, *Chief of the Imperial General Staff, December 1915–February 1918* (London, 1989), 50, 60, 104–5; David R. Woodward, *Lloyd George and the Generals* (Newark, 1983), 107–13.

56. Blake, *Haig*, 44–5, 203–9.

57. John Turner, *British Politics and the Great War: Coalition and Conflict 1915–1918* (New Haven, 1992), 161.

58. Blake, *Haig*, 344.

59. David Fraser, *Alanbrooke* (London, 1982), 395; Brian Bond (ed.), *Chief of Staff: The Diaries of Lieutenant-General Sir Henry Pownall* (2 vols., London, 1972–4), i. 258, 264–5, 358.

60. Nigel Hamilton, *Monty: The Field Marshal 1944–1976* (London, 1986), 670.

61. Antony Beevor, 'On Her Majesty's Depleted Service', *Spectator* (25 Jan. 1992), 14–15.

62. Peter Dietz, *The Last of the Regiments: Their Rise and Fall* (London, 1990), 26, 29. For a statement of the position to which Dietz referred, without his reservations, see John Baynes, *No Reward but Honour? The British Soldier in the 1990s* (London, 1991), 61.

63. Clode, *Military Forces of the Crown*, i. 29–30; see also i, pp. v–vi, 280; ii. 392–429.

Notes to Chapter 4

1. For examples of this sort of argument, see Otley, 'Militarism and the Social Affiliations of the British Army Élite', 87; John Keegan, 'Western Europe and its Armies 1945–85', in L. H. Gann (ed.), *The Defense of Western Europe* (London, 1987), 9.

2. An English translation of Lyautey's article, by W. K. West, was published in *Army Quarterly*, 119 (1980), 77–94.

3. Philip Stigger, ' Major L. F. Knollys, C.M.G., Professional Aide-de-Camp and Colonial Military Policeman', *Journal of the Society for Army Historical Research*, 73 (1995), 50.

4. M. E. Yapp, *Strategies of British India: Britain, Iran and Afghanistan 1798–1850* (Oxford, 1980), 442.

5. Douglas M. Peers, 'Between Mars and Mammon: The East India Company and Efforts to Reform its Army, 1796–1832', *Historical Journal*, 33 (1990), 394, 396. Peers's article is important for much of what follows.

6. C. A. Bayly, *Imperial Meridian: The British Empire and the World 1780–1830* (London, 1989), 134–5.

7. Edward Ingram, *In Defence of British India: Great Britain in the Middle East, 1775–1842* (London, 1984), 50.

8. C. A. Philips (ed.), *The Correspondence of Lord William Cavendish Bentinck, Governor-General of India 1828–1835* (2 vols., Oxford, 1977), esp. pp. xix–xx, xlii–iv, 264, 298; John Rosselli, *Lord William Bentinck: The Making of a Liberal Imperialist 1774–1839* (London, 1974), 45–6, 50–4, 309–15.

9. Douglas M. Peers, 'Soldiers, Scholars, and the Scottish Enlightenment:

Militarism in Early Nineteenth-Century India', *International History Review,* 16 (1994), 441–65; also Peers, 'Between Mars and Mammon', 386–8.

10. Yapp, *Strategies of British India,* esp. 9–10, 588–91; see also Ingram, *In Defence of British India,* 180–4.

11. W. Murray Hogben, 'British Civil–Military Relations on the North-West Frontier of India', in Adrian Preston and Peter Dennis (eds.), *Swords and Covenants* (London, 1976); especially revealing is Philip Mason's anecdote, retold on p. 123.

12. Ibid. 139.

13. Priscilla Napier, *Revolution and the Napier Brothers 1820–1840* (London, 1973), 223.

14. W. F. P. Napier, *The Life and Opinions of General Sir Charles James Napier GCB* (4 vols., London, 1857), i. 230–1.

15. Ibid. 470.

16. Ibid. 323.

17. Major-General Sir Charles Napier (ed.), *Lights and Shades of Military Life* (2 vols., London, 1840), i. 329–32.

18. Ibid., pp. vii, 381–3.

19. Ibid. 329; see also Napier, *Life and Opinions of Charles Napier,* ii. 121.

20. Napier, *Lights and Shades of Military Life,* ii. 323–4.

21. Outram to his mother, Nov. 1833, quoted in Lionel J. Trotter, *The Bayard of India: A Life of General Sir James Outram* (London, 1909), 32.

22. The fullest recent accounts of the Napier/Outram feud are Priscilla Napier, *I have Sind: Charles Napier in India 1841–1844* (Salisbury, 1990), and *Raven Castle: Charles Napier in India 1844–1851* (Salisbury, 1991). See also Yapp, *Strategies of British India,* 262, 439, 472–5, 494–5.

23. Napier, *Life and Opinions of Charles Napier,* iv. 19–20.

24. Napier, *Raven Castle,* 201.

25. Ibid. 198.

26. J. G. A. Baird (ed.), *Private Letters of the Marquess of Dalhousie* (Edinburgh, 1910), 123.

27. Napier, *Raven Castle,* 220.

28. Baird, *Private Letters of Dalhousie,* 127.

29. Lieutenant-General Sir Charles James Napier, *Defects, Civil and Military, of the Indian Government* (4th edn. London, 1857), 221–3.

30. Ibid. 221.

Notes to Chapter 5

1. Brian Robson, *The Road to Kabul: The Second Afghan War 1878–1881* (London, 1986), 250–1.

2. Robson, *Roberts in India,* 23.

3. Ibid. 47–8, 54, 97–8.

4. Ibid. 119–20.

5. Ibid., pp. xix–xxi; Robson, *Road to Kabul,* 139–44.

6. Robson, *Roberts in India,* 150–2, 186–94.

7. Ibid. 206–14.
8. Ibid. 157, 186.
9. G. J. Wolseley, 'Long and Short Service', *Nineteenth Century*, 9 (Mar. 1881), 560.
10. Symons, *Buller's Campaign*, 11.
11. Adrian Preston, 'Wolseley, the Khartoum Relief Expedition and the Defence of India, 1885–1900', *Journal of Imperial and Commonwealth History*, 6 (1978), 254–80. This essay is also to be found in Preston and Dennis, *Swords and Covenants*.
12. Adrian Preston (ed.), *In Relief of Gordon: Lord Wolseley's Campaign Journal of the Khartoum Relief Expedition* (London, 1967), pp. xxxix, 80, 136.
13. Quoted in Le May, *Victorian Constitution*, 80.
14. See especially Adrian Preston, 'Introduction', in Peter Dennis and Adrian Preston, *Soldiers as Statesmen* (London, 1976), 31–3.
15. Sir F. Maurice and Sir George Arthur, *The Life of Lord Wolseley* (London, 1924), 268.
16. John Wilson, *CB: A Life of Sir Henry Campbell-Bannerman* (London, 1973), 205.
17. Hansard, 3rd series, cccxxvi, cols. 103–7, 14 May 1888.
18. Ibid., col. 91.
19. Quoted in Spiers, *The Army and Society*, 227.
20. General Viscount Wolseley, *The Life of John Churchill, Duke of Marlborough to the Accession of Queen Anne* (2 vols., London, 1894), ii. 2–4, 12, 41.
21. Ibid. 83.
22. Ibid. 84–5.
23. Ibid. 434; also 88.
24. Ibid. 434.
25. Ibid. 3, 68, 71, 81, 435.
26. Henderson, *Stonewall Jackson*, i, p. x.
27. David Gilmour, *Curzon* (London, 1994), 251.
28. Hansard, 4th series, xc, cols. 341–2, 4 Mar. 1901.
29. Dilks, *Curzon*, ii. 107. Dilks is the fullest guide to the Kitchener–Curzon struggle, but see also Gilmour, *Curzon*, and George Cassar, *Kitchener: Architect of Victory* (London, 1977).
30. Gilmour, *Curzon*, 329.
31. Cassar, *Kitchener*, 149, 150.
32. Dilks, *Curzon*, ii. 107.
33. Ibid. 216.
34. Cassar, *Kitchener*, 153.
35. Gilmour, *Curzon*, 337.
36. Dilks, *Curzon*, ii. 207.
37. Information from Dr Peter Martland.
38. Williams, *Defending the Empire*, 55–8, 142–55, 180–225; R. J. Q. Adams and Philip P. Poirier, *The Conscription Controversy in Great Britain, 1900–18* (London, 1987), 1–47.
39. Adams and Poirier, *Conscription Controversy*, 26.

40. Williams, *Defending the Empire*, 188.

41. Field Marshal Earl Roberts, *Fallacies and Facts: An Answer to 'Compulsory Service'* (London, 1911), esp. 5–8; see also General Sir Ian Hamilton, *Compulsory Service: A Study of the Question in the Light of Experience*, with an introduction by Viscount Haldane of Cloan (London, 1910).

42. C. E. Callwell, *Field Marshal Sir Henry Wilson: His Life and Diaries* (2 vols., London, 1927), i. 76–7; Brian Bond, *The Victorian Army and the Staff College* (London, 1972), 262–3.

43. Elizabeth A. Muenger, *The British Military Dilemma in Ireland: Occupation Politics, 1886–1914* (Lawrence, Kan., 1991), 165–6.

44. Robert Blake, 'Great Britain: The Crimean War to the First World War', in Howard, *Soldiers and Governments*, 37.

45. I am grateful to Jonathan Frew for this figure.

46. Beckett, *The Army and the Curragh Incident, 1914*, in addition to providing a splendid collection of documents has a most helpful introduction; Sir James Fergusson, *The Curragh Incident* (London, 1964), is the best general account, written by the son of a key participant.

47. Muenger, *British Military Dilemma in Ireland*, 191.

48. Callwell, *Wilson*, i. 47.

49. Beckett, *The Army and the Curragh Incident*, 281.

50. Ibid. 53.

51. Ibid. 205–7, 209–10.

52. Ibid. 142.

53. Ibid. 55.

Notes to Chapter 6

1. Clode, *Military Forces of the Crown*, i. 275–7; ii. 200–2, 274, 322–4, 392–429.

2. Ibid. ii. 411–18.

3. Parliamentary papers 1887, Command 5062, p. 100; Hamer, *British Army*, 119; Hamer is the best source for what follows.

4. C. W. Dilke and Spenser Wilkinson, *Imperial Defence* (London, 1897; 1st edn. 1892), 169; see also Spenser Wilkinson, 'The General Management of the Army', *Nineteenth Century*, 47 (Feb. 1900), 186.

5. John Kennedy, *The Business of War: The War Narrative of Major-General Sir John Kennedy G.C.M.G., K.C.V.O., K.B.E., C.B., M.C.* (London, 1957), 14.

6. John Ehrman, *Cabinet Government and War 1890–1940* (Cambridge, 1958), 29–31, 41–3; see also Nicholas d'Ombrain, *War Machinery and High Policy: Defence Administration in Peacetime Britain 1902–1914* (London, 1973).

7. Franklyn Arthur Johnson, *Defence by Committee: The British Committee of Imperial Defence 1885–1959* (London, 1960), 131.

8. C. R. M. F. Cruttwell, *A History of the Great War* (2nd edn. Oxford, 1936), 501.

9. Bond, *The Victorian Army and the Staff College*, 263–6; Bassford, *Clausewitz in English*, 75–6; W. N. Nicholson, *Behind the Lines* (London, 1939), 169.

10. Abrams, 'Democracy, Technology and the Retired British Officer', 161.

11. David French, 'The Rise and Fall of "Business as Usual"', in Kathleen Burk (ed.), *War and the State: The Transformation of British Government 1914–1919* (London, 1982), 17–18.

12. Avner Offer, *The First World War: An Agrarian Interpretation* (Oxford, 1989), 293–6, 300–10; Arthur J. Marder, *From the Dreadnought to Scapa Flow: The Royal Navy in the Fisher Era, 1904–1919* (5 vols., London, 1961–70), i. 342–4.

13. Ehrman, *Cabinet Government and War*, 56–8; Lord Hankey, *Government Control in War* (Cambridge, 1945), 36.

14. John Gooch, *The Plans of War: The General Staff and British Military Strategy c.1900–1916* (London, 1974), 299–315.

15. A representative bibliography of the events of May 1915 would include the following: Cameron Hazlehurst, *Politicians at War July 1914 to May 1915: A Prologue to the Triumph of Lloyd George* (London, 1971); Trevor Wilson, *The Downfall of the Liberal Party 1914–1935* (London, 1966); Stephen Koss, *Asquith* (London, 1976); Stephen Koss, 'The Destruction of Britain's Last Liberal Government', *Journal of Modern History*, 40 (1968), 257–77; Peter Fraser, 'British War Policy and the Crisis of Liberalism in May 1915', *Journal of Modern History*, 54 (1982), 1–26; Martin D. Pugh, 'Asquith, Bonar Law and the First Coalition', *Historical Journal*, 17 (1974), 813–36.

16. David French, 'The Meaning of Attrition, 1914–1916', *English Historical Review*, 103 (1988), 385–405.

17. George H. Cassar, *The Tragedy of Sir John French* (Newark, 1985), 199–200.

18. Callwell, *Wilson*, i. 200–1; Gwynne to Wilson, 12 Apr. 1915, in Keith Wilson (ed.), *The Rasp of War: The Letters of H. A. Gwynne to the Countess Bathurst 1914–1918* (London, 1988), 76–7.

19. Cassar, *Tragedy of Sir John French*, 239.

20. John Grigg, *Lloyd George: From Peace to War 1912–1916* (London, 1985), 244.

21. Stephen Koss has argued that Churchill conspired with French and Repington in a visit to GHQ on 8 and 9 May (*Asquith*, 193–4). This seems improbable, given both French's preoccupation with the Aubers ridge attack on those days, and the likelihood that Churchill would be a casualty rather than a beneficiary of what would follow.

22. Grigg, *Lloyd George*, 244; Gerard J. De Groot, *Douglas Haig, 1861–1928* (London, 1988), 190.

23. Hazlehurst, *Politicians at War*, 238.

24. Field Marshal Viscount French of Ypres, *1914* (London, 1919), 357, 361.

25. Wilson, *The Rasp of War*, 153–4; Blake, *Haig*, 168, 254.

26. Woodward, *Lloyd George and the Generals*, 78–81; Woodward (ed.), *Military Correspondence of Robertson*, 10–12.

27. A. D. Harvey, *Collision of Empires: Britain in Three World Wars, 1793–1945* (London, 1992), 471.

28. Adams and Poirier, *Conscription Controversy*, 149.

29. Keith Neilson, *Strategy and Supply: The Anglo-Russian Alliance, 1914–17* (London, 1984), 281.

30. Keith Jeffery, *The British Army and the Crisis of Empire, 1918–22* (Manchester, 1984), 81.

31. Trevor Wilson, *The Myriad Faces of War: Britain and the Great War 1914–1918* (Cambridge, 1986), 463–4.

32. Woodward (ed.), *Military Correspondence of Robertson*, 346.

33. Brock Millman, 'Henry Wilson's Mischief: Field Marshall [*sic*] Sir Henry Wilson's Rise to Power 1917–18', *Canadian Journal of History*, 30 (1995), 467–86.

34. Wilson, *The Rasp of War*, 246–7.

35. Nancy Maurice (ed.), *The Maurice Case: From the Papers of Major-General Sir Frederick Maurice K.C.M.G., C.B.* (London, 1972), 60, 98. See also John Gooch, 'The Maurice Case', in *The Prospect of War: Studies in British Defence Policy 1847–1942* (London, 1981), and David R. Woodward, 'Did Lloyd George Starve the British Army of Men prior to the German Offensive of 21 March 1918', *Historical Journal*, 27 (1984), 241–52.

36. Maurice, *The Maurice Case*, 118.

37. Wilson, *The Rasp of War*, 274–8.

38. Edward David, 'The Liberal Party Divided 1916–1918', *Historical Journal*, 13 (1970), 518–20; Wilson, *Downfall of the Liberal Party*, 110–12; Robert Blake, *The Unknown Prime Minister: The Life and Times of Andrew Bonar Law 1858–1923* (London, 1955), 370–2; John Stubbs, 'The Impact of the Great War on the Conservative Party', in Gillian Peele and Chris Cook (eds.), *The Politics of Reappraisal 1918–1939* (London, 1975), 30.

39. Keith Jeffery (ed.), *The Military Correspondence of Field Marshal Sir Henry Wilson 1918–1922* (London, 1985), 43.

40. Blake, *Haig*, 308.

41. These points are developed in Hew Strachan, 'Germany in the First World War: The Problem of Strategy', *German History*, 12 (1994), 237–49.

42. Keith Grieves, 'C. E. Montague and the Making of *Disenchantment*, 1914–1921', *War in History*, 4 (1997); see also F. Maurice, *Governments and War: A Study of the Conduct of War* (London, 1926), 14–15.

Notes for Chapter 7

1. See the essays in Brian Bond (ed.), *The First World War and British Military History* (Oxford, 1991), esp. Hew Strachan, '"The Real War": Liddell Hart, Cruttwell, and Falls', and Ian Beckett, 'Frocks and Brasshats'.

2. Sir William Robertson, *Soldiers and Statesmen 1914–1918* (2 vols., London, 1926), i. p. xiii.

3. Ibid. 153, 167.

4. Ibid. 180, 189.

5. Ibid. ii. 299–304.

6. Maurice, *The Maurice Case*, 181; see also 158–62.

7. Maurice, *Governments and War*, 8, 150–1.

8. Ibid. 44.

9. Ibid. 13.

10. Ibid. 121–3; also 128.

11. Ibid. 127.

12. Ibid. 156.

13. Ibid. 112; also 97–8, 104.

14. Ibid. 53, 76–7.

15. Maurice, *The Maurice Case*, 115.

16. Maurice, *Governments and War*, 139.

17. Hankey, *Government Control in War*, 55; Johnson, *Defence by Committee*, 192–7.

18. Johnson, *Defence by Committee*, 174.

19. Ibid. 233; Maurice, *Governments and War*, 143, 146–7, 158, 166–71.

20. Kennedy, *The Business of War*, 254.

21. *The Memoirs of General the Lord Ismay, K.G., P.C., G.C.B., C.H., D.S.O.* (London, 1960), 16, 39–40, 50, 167.

22. J. R. Colville, *Man of Valour: The Life of Field Marshal the Viscount Gort, V.C., G.C.B., D.S.O., M.V.O., M.C.* (London, 1972), 80–3; for what follows, see especially A. J. Trythall, 'The Downfall of Leslie Hore-Belisha', *Journal of Contemporary History*, 16 (1981), 391–411; Brian Bond, 'Leslie Hore-Belisha at the War Office', in Ian Beckett and John Gooch (eds.), *Politicians and Defence: Studies in the Formulation of British Defence Policy 1845–1970* (Manchester, 1981); Bond, *Chief of Staff*. The account in R. J. Minney, *The Private Papers of Hore-Belisha* (London, 1960) is not very revealing.

23. J. P. Harris, 'Two War Ministers: A Reassessment of Duff Cooper and Hore-Belisha', *War & Society*, 6 (1988), 69–73.

24. Bond, *Chief of Staff*, i. 206; also 130–2, 136, 147–9, 202–3, 208–10, 215.

25. Ibid. 209.

26. Colville, *Man of Valour*, 135.

27. Bond, *Chief of Staff*, i. 209–10.

28. Kennedy, *The Business of War*, 34.

29. Trythall, 'Downfall of Hore-Belisha', 403.

30. Bond, *Chief of Staff*, i. 254.

31. *The Ironside Diaries 1937–40*, ed. Colonel Roderick Macleod and Denis Kelly (London, 1962), 166–7; also 105, 125.

32. Bond, *Chief of Staff*, i. 262–4.

33. *Ironside Diaries*, 194; Kennedy, *The Business of War*, 151. Both Trythall and Bond seem doubtful about calling what occurred a conspiracy, but that in its turn rests on minimizing the role of Pownall.

34. Trythall, 'Downfall of Hore-Belisha', 401.

35. Bond, *Chief of Staff*, i. 258.

36. Ibid. 265.

37. Colville, *Man of Valour*, 162.

38. Bond, *Chief of Staff*, i. 268.

39. Ibid. i. 274.

40. Trythall, 'Downfall of Hore-Belisha', 406.

41. Arthur Bryant, *The Turn of the Tide 1939–1943* (London, 1957), 21.

42. For a general discussion of this and related points, see Alex Danchev,

'Waltzing with Winston: Civil–Military Relations in Britain in the Second World War', *War in History*, 2 (1995), 202–30.

43. Bond, *Chief of Staff*, i. 282.
44. Kennedy, *The Business of War*, 114–15; also Bryant, *Turn of the Tide*, 337.
45. Fraser, *Alanbrooke*, 440, 442.
46. Alex Danchev, ' "Dilly-Dally", or Having the Last Word: Field Marshal Sir John Dill and Prime Minister Winston Churchill', *Journal of Contemporary History*, 22 (1987), 21–44; Danchev, 'Waltzing with Winston', 207.
47. Kennedy, *Business of War*, 63–4.
48. Ibid. 119–20, 128–9, 143.
49. Danchev, 'Waltzing with Winston', 203–4.
50. Bryant, *Turn of the Tide*, 626–7.
51. Fraser, *Alanbrooke*, 416–21.
52. See, for example, Bond, *Chief of Staff*, ii. 33–4; Kennedy, *The Business of War*, 97, 256.
53. Kennedy, *The Business of War*, 245.
54. Fraser, *Alanbrooke*, 442.
55. On the fashioning of this consensus, see Danchev, ' "Dilly-Dally" ', 29–33. More generally, see also John Keegan, 'Churchill's Strategy', and Michael Carver, 'Churchill and the Defence Chiefs', in Robert Blake and Wm. Roger Louis (eds.), *Churchill* (Oxford, 1993).

Notes to Chapter 8

1. Three general texts on Britain's experience of counter-insurgency are Charles Townshend, *Britain's Civil Wars: Counterinsurgency in the Twentieth Century* (London, 1986); Thomas R. Mockaitis, *British Counterinsurgency, 1916–60* (London, 1990); David A. Charters, 'From Palestine to Northern Ireland: British Adaptation to Low-Intensity Operations', in D. A. Charters and Maurice Tugwell (eds.), *Armies in Low-Intensity Conflict: A Comparative Analysis* (London, 1989). See also T. R. Mockaitis, 'The Origins of British Counter-insurgency', *Small Wars and Insurgencies*, 1 (1990), 209–25, and T. R. Mockaitis, *British Counterinsurgency in the Post-imperial Era* (Manchester, 1995).
2. Townshend, *Britain's Civil Wars*, 61.
3. Colm Campbell, *Emergency Law in Ireland, 1918–1925* (Oxford, 1994), 134–5, 144–5.
4. Charles Townshend, *The British Campaign in Ireland 1919–1921: The Development of Political and Military Policies* (Oxford, 1975), esp. 13, 46, 74–9, 81–2, 94–5, 103–5, 110–12, 135–7, 160–2, 173, 184.
5. Alfred Draper, *The Amritsar Massacre: Twilight of the Raj* (Leatherhead, 1985; 1st pub. 1981), 31.
6. Ibid. 98.
7. Ibid. 155.
8. Ibid. 222.
9. Jeffery (ed.), *Military Correspondence of Sir Henry Wilson*, 184–7; Callwell, *Wilson*, ii. 237–9.

10. Charles Gwynn, *Imperial Policy* (London, 1936; 1st edn. 1934), 16.
11. H. J. Simson, *British Rule, and Rebellion* (Edinburgh, 1937), esp. 80, 117.
12. Charters, 'From Palestine to Northern Ireland', 189–90.
13. David A. Charters, *The British Army and Jewish Insurgency in Palestine, 1945–47* (London, 1989), esp. 85–90, 100–10, 125–7, 134–62, 174–6.
14. Townshend, *Britain's Civil Wars*, 158.
15. Anthony Short, *The Communist Insurrection in Malaya 1948–1960* (London, 1975), 334.
16. Ibid. 386.
17. Mockaitis, *British Counterinsurgency*, 135; see Townshend, *Britain's Civil Wars*, 27.
18. Julian Paget, *Counter-insurgency Campaigning* (London, 1967), 157–9.
19. Tom Pocock, *Fighting General: The Public and Private Campaigns of General Sir Walter Walker* (London, 1973); see also Raffi Gregorian, 'Jungle Bashing in Malaya: Towards a Formal Tactical Doctrine', *Small Wars and Insurgencies*, 5 (1994), 338–59.
20. Mockaitis, *British Counterinsurgency in the Post-imperial Era*, 15–16.
21. Pocock, *Fighting General*, 155–99.
22. Ibid. 219–39.
23. Ibid. 248.
24. The evidence for this is largely anecdotal. Some officers serving in 1965 are categorical that they were consulted on these issues; some equally clearly state they were not. For the memoir literature of politicians, see Adam Roberts, 'The British Armed Forces and Politics: A Historical Perspective', *Armed Forces and Society*, 3 (1977), 541, and n. 23.
25. Walter Walker, *The Bear at the Back Door: The Soviet Threat to the West's Lifeline in Africa* (Richmond, 1978), 23; also 18.
26. Ibid. 56; Michael Carver, *Out of Step: The Memoirs of Field Marshal Lord Carver* (London, 1989), 502–4.
27. Walker, *The Bear at the Back Door*, pp. xvii, 83.
28. Walter Walker, *The Next Domino?* (London, 1980), 150.
29. Walker, *The Bear at the Back Door*, 185.
30. Ibid. 17–18, 179, 211–22; Walker, *The Next Domino?*, 285.
31. Walker, *The Bear at the Back Door*, 164.
32. Ibid. 179.
33. Ibid. 180–4; Walker, *The Next Domino?*, 267.
34. Walker, *The Next Domino?*, 259–60.
35. Desmond Hamill, *Pig in the Middle: The Army in Northern Ireland 1969–1984* (London, 1985), 83–4.
36. Clive Walker, 'The Role and Powers of the Army in Northern Ireland', in Brigid Hadfield (ed.), *Northern Ireland: Politics and the Constitution* (Buckingham, 1992); also Townshend, *Britain's Civil Wars*, 69–71.
37. Walker, 'The Role and Powers of the Army in Northern Ireland', 120, 121.
38. Mockaitis, *British Counterinsurgency in the Post-imperial Era*, 127.
39. Baynes, *No Reward but Honour?*, 72.

40. John Newsinger, 'From Counter-insurgency to Internal Security: Northern Ireland 1969–1992', *Small Wars and Insurgencies*, 6 (1995), 88–111.
41. Walker, 'The Role and Powers of the Army in Northern Ireland'; on Clegg, see esp. Roy Greenslade, 'How the Press Freed Lee Clegg', *The Times*, 5 July 1995.
42. Frank Kitson, *Low Intensity Operations: Subversion, Insurgency, Peace-keeping* (London, 1971), 170, 172, 52.
43. Ibid. 69–70.
44. Ibid. 169.
45. Ibid. 13.
46. Ibid. 24–5.
47. Roger Faligot, *Britain's Military Strategy in Ireland: The Kitson Experiment* (Dingle, 1983; 1st pub. in France, 1980), 1; for a different and more balanced view see Bruce Page and Lewis Chester, 'The Guru of the New Model Army', *Sunday Times*, 14 May 1972.
48. Tony Geraghty, *Who Dares Wins: The Story of the Special Air Service 1950–1980* (London, 1980), esp. 16, 18, 164–6, 170–3.
49. Christopher J. Whelan, 'Armed Forces, Industrial Disputes and the Law in Britain', in Peter J. Rowe and Christopher J. Whelan (eds.), *Military Intervention in Democratic Societies* (London, 1985), 110–11; see also the other essays in the same volume; Richard Norton-Taylor, 'Any Help the Army could Currently Provide would be Extremely Limited', *Guardian*, 19 Jan. 1979; Peter Hennessy, 'What the Military Thinks about its Role', *The Times*, 19 Nov. 1979.
50. Jack Woddis, *Armies and Politics* (London, 1977), 275.
51. Ibid. 29.
52. Roberts, 'The British Armed Forces and Politics', 532.
53. Henry Stanhope, 'The Army is more Interested in Rugby than Revolution', *The Times*, 17 Sept. 1974.
54. Sir Edwin Bramall, 'The Place of the British Army in Public Order', in Rowe and Whelan, *Military Intervention in Democratic Societies*, 82.
55. Ibid. 77–9; also Sir Robert Mark, 'Keeping the Peace in Great Britain: The Differing Roles of the Police and the Army', ibid. 87–9; General Sir Hugh Beach, 'Internal Uses of the Military', lecture at RMA Sandhurst, 16 Jan. 1979.
56. Frank Kitson, *Bunch of Five* (London, 1977), p. xii.
57. Ibid. 283–300; Frank Kitson, *Directing Operations* (London, 1989), 50–7.
58. Kitson, *Directing Operations*, 117; also 21–2, 43–4, 111–18.

Notes to Chapter 9

1. Patrick Mileham, 'Ethos: British Army Officership 1962–1992', *Strategic and Combat Studies Institute Occasional Papers*, 19 (1995), 22.
2. Brigadier A. G. Denaro, 'Warrior or Worker: Is the British Army Producing the Right Men to Command its Troops on Operations?' *Journal of the Royal United Services Institute*, 140 (June 1995), 40–1.

3. Zugbach, *Power and Prestige*, 2–5.

4. John Keegan, 'Western Europe and its Armies 1945–85', in Gann, *The Defense of Western Europe*, 7–9; John Keegan's best-known discussion of the regimental system, derived from his Sandhurst experiences, is 'Regimental Ideology', in Geoffrey Best and Andrew Wheatcroft (eds.), *War, Economy and the Military Mind* (London, 1976). See also David Weston, 'The Army: Mother, Sister and Mistress: The British Regiment', in Edmonds, *The Defence Equation*.

5. Michael Howard, 'Soldiers in Politics', *Encounter*, 19 (Sept. 1962), 77–81.

6. Zugbach, *Power and Prestige*, 26–8, 173–7.

7. J. W. Fortescue, *A History of the British Army* (13 vols., London, 1899–1930), xiii. 576–7.

8. Guy, *Oeconomy and Discipline*, 40–2.

9. Wolseley, 'Long and Short Service', 558–9, 572.

10. Parliamentary papers, 1881, XXI, *Report of Committee . . . to Inquire into Militia and the Present Brigade Depot System* (1877).

11. Ibid., *Report of a Committee of General and Other Officers of the Army on an Army Re-organization* (1881), 312.

12. Maurice and Arthur, *Life of Lord Wolseley*, 136–7.

13. Hansard, 3rd series, cclxi, cols. 518–43, 16 May 1881.

14. Verner, *Duke of Cambridge*, ii. 202.

15. Ibid. ii. 213–16.

16. Childers, *Life and Correspondence of Hugh Childers*, ii. 49–50.

17. H. J. Hanham, 'Religion and Nationality in the Mid-Victorian Army', in M. R. D. Foot (ed.), *War and Society* (London, 1973), 163.

18. WO 33/36, pp. 114–31.

19. Alan Ramsay Skelley, *The Victorian Army at Home* (London, 1977), 289–94.

20. General the Rt. Hon. Sir Nevil Macready, *Annals of an Active Life* (2 vols., London, 1924), 257.

21. P. E. Dewey, 'Military Recruiting and the British Labour Force during the First World War', *Historical Journal*, 27 (1984), 199–223.

22. Nicholas Perry, 'Nationality in the Irish Infantry Regiments in the First World War', *War & Society*, 12 (1994), 65–95; Hansard, 5th series, lxxxvi, cols. 646–52, 18 Oct. 1916.

23. Fortescue, *History of the British Army*, xiii. 572.

24. Perry, 'Nationality', 72.

25. Edward Spiers, 'The Scottish Soldier at War', in Hugh Cecil and Peter Liddle (eds.), *Facing Armageddon: the First World War Experienced* (Barnsley, 1996).

26. WO 32/5959, proof Army Order, 15 Mar. 1921.

27. WO 32/5959, French to Secretary of State for War, 15 Mar. 1921.

28. WO 32/5960, Cavan to Wilson, 31 Jan. 1922.

29. WO 32/5960, memorandum by Kirke, 30 July 1921.

30. Captain J. R. Kennedy, *This, our Army* (London, 1935), 75–90.

31. I am grateful to Dr S. D. Badsey for this point; see his article 'Faction in

the British Army: Its Impact on 21st Army Group Operations in Autumn 1944', *War Studies Journal*, 1 (1995), 13–28, esp. 15.

32. Brian Bond, *British Military Policy between the Two World Wars* (Oxford, 1980), 62; Correlli Barnett, *The Desert Generals* (London, 1960), 99–105; Shelford Bidwell and Dominick Graham, *Fire-Power: British Army Weapons and Theories of War 1904–1945* (London, 1982), 190–1, 193.

33. Craig Forsyth French, 'The 51st Highland Division from St Valery to El Alamein', M.Phil. thesis (Glasgow University, 1996).

34. Quoted by Saul David, *Mutiny at Salerno: An Injustice Exposed* (London, 1995), 148; see also 35–7, 41–50, 138–9; Lawrence James, *Mutiny in the British and Commonwealth Forces 1797–1956* (London, 1987), 171–3; A. Patient, *Listener*, 25 Feb. 1982.

35. W. A. Elliott, *Esprit de Corps: A Scots Guards Officer on Active Service 1943–1945* (Norwich, 1996), 104–6, 109–10.

36. David Fraser, *And We Shall Shock Them: The British Army in the Second World War* (London, 1983), 87, 92; Fraser, *Alanbrooke*, 216.

37. Bond, *Chief of Staff*, ii. 98.

38. Gregory Blaxland, *The Regiments Depart: A History of the British Army 1945–1970* (London, 1971), 329–30.

39. John Cloake, *Templer: Tiger of Malaya* (London, 1985), 37–8, 379.

40. WO 163/573, Report of the Committee on the Organization of the Army (Sept. 1956), 10.

41. Hansard, 5th series, 566, cols. 838, 892, 11 Mar. 1957; 570, cols. 269, 284, 324, 326–7, 339, 359, 14 May 1957; 579, cols. 1243–6, 11 Dec. 1957; 592, written answer, col. 200, 31 July 1958.

42. Ibid. 579, cols. 1247–51, 11 Dec. 1957.

43. Pocock, *Fighting General*, 119; for what follows see 115–26, 131, 142–51.

44. Lt.-Col. J. H. Jacob to the editor, *The Times*, 10 July 1968.

45. Colin Mitchell, *Having Been a Soldier* (London, 1969).

46. *The Times*, 16, 17, 18 July 1968; *Scotsman*, 17, 18 July 1968; for Denis Healey's written reply, *The Times* and *Scotsman*, 25 July 1968.

47. *Scotsman*, 22 and 27 July 1968, 6 Aug. 1968.

48. *Punch*, 15 July 1970, p. 90.

49. *Scottish Daily Express*, 9 July 1968; *The Times*, 16 July 1968.

50. *Scotsman*, 27 July 1968.

51. *The Times*, 13 July 1968.

52. Alastair F. Lamont to the editor, *Scotsman*, 23 July 1968; see also Alastair I. R. Stewart to the editor, *Scotsman*, 18 July 1968.

53. *The Times*, 12 July 1968.

54. Michael Yardley and Dennis Sewell, *A New Model Army* (London, 1989), 45.

55. Ibid. 45; Blaxland, *The Regiments Depart*, 472.

56. Frank Kitson, *Warfare as a Whole* (London, 1987), 87, 89, 147–9.

57. For example, *Daily Telegraph*, 28 Jan. 1988; Grahame Keating and William F. Owen, 'Back to the Future: Britain's New Royal Flying Corps', *Journal of the Royal United Services Institute*, 141 (Apr. 1996), 28–33.

58. Captain J. P. Riley, 'MCV-80 and beyond: Implications for the Infantry', *Journal of the Royal United Services Institute*, 131 (Sept. 1986), 28; Captain R. A. D. Applegate, 'Why Armies Lose in Battle: An Organic Approach to Military Analysis', ibid. 132 (1987), 52–3; see also *The Times*, 18 Mar. 1983.

59. Antony Beevor, *Inside the British Army* (London, 1990), pp. xxvii, 308.

60. Zugbach, *Power and Prestige*, 9, 106.

61. Beevor, *Inside the British Army*, 127–34.

62. Antony Beevor, 'On Her Majesty's Depleted Service', *Spectator*, 25 Jan. 1992, pp. 14–15.

63. P. A. J. Cordingley, 'The Gulf War: Operating with Allies', *Journal of the Royal United Services Institute for Defence Studies*, 136 (Apr. 1991), 17; General Sir Peter de la Billière, *Storm Command: A Personal Account of the Gulf War* (London, 1992), 170, 191.

64. See evidence of Tom King to *House of Commons Defence Committee: 3rd Report: Options for Change: Army* (1991–2), HC 45, p. 83.

65. The papers of the KOSB Campaign have been deposited in the National Library of Scotland (hereafter NLS), manuscripts Acc 11230; a published account, from the perspective of the King's Own Scottish Borderers, is Donald Fairgrieve, *A Regiment Saved: An Account of Operation Borderer 23rd July 1991 to 3rd February 1993* (Edinburgh, 1993).

66. See the exchange of letters between General Sir Gordon MacMillan, Major-General F. C. C. Graham, and Malcolm MacPherson, MP, *Scotsman*, 22 Feb., 27 Feb., 1 Mar., 6 Mar. 1969; also MacMillan to the editor, *The Times*, 25 June 1970.

67. NLS Acc 11230/1, Hamilton to D. A. Robson, 14 Oct. 1991; *Scotsman*, 27 Jan. 1992, quoted in *House of Commons Defence Committee: 3rd Report* (1991–2), HC 45, p. 89.

68. NLS Acc 11230/1–2, MacMillan to Inge, 18 June 1992; MacMillan to all Scottish MPs, 20 June 1992; 11230/3, MacMillan to Chapple, 25 Aug. 1991; 11230/4, MacMillan to Hamish [?Logan], 5 June 1992; memo of 8 July 1992 for meeting of 21 Aug. 1992; 11230/5, minutes of meetings, 21 Aug. 1992; Citigate report, 25 May 1993; position paper by Michael Ashmore, 18 Oct. 1993; minutes of meeting, 4 Feb. 1994; 11230/6, minutes of meeting, 29 Apr. 1994; Fairgrieve, *A Regiment Saved*, 50–1, 70, 95–6, 109–10, 113.

69. NLS Acc 11230/1–2, Mark Tennant, 23 Mar. 1992.

70. Neil R. MacCallum to the editor, *Scotsman*, 26 Sept. 1994.

71. See the suggestive letter from Lady MacMillan to the editor, *The Times*, 18 Oct. 1991.

72. NLS Acc 11230/1, Citigate's plan, Aug. 1991.

73. NLS Acc 11230/7.

74. D. A. Robson to the editor, *Sunday Times Scotland*, 12 Feb. 1995; *The Times*, 22 Oct. and 25 Oct. 1991; *Scotsman*, 26 Oct. 1991; also *Scotsman*, 12 and 15 Aug. 1991.

75. NLS Acc 11230/1–2, referring to *Scotsman*, 27 Jan. 1992.

76. Ibid., Paul Martin, 16 Sept. 1991; W. D. Fairgrieve, 3 Feb. 1992; D. A.

Robson to Charles Laidlaw, 2 Mar. 1992; 11230/4, minutes of meeting, 9 Jan. 1992.

77. Fairgrieve, *A Regiment Saved*, 91–2; NLS Acc 11230/1–2, MacMillan to all Scottish MPs, 20 June 1992.

78. NLS Acc 11230/1–2, MacMillan to the prime minister, 27 Mar. 1992.

79. *Bulletin of the Military Historical Society*, 45 (1994), 105.

80. NLS Acc 11230/5, minutes of meeting, 10 Sept. 1993; strengths of Scottish Division 1987-93, annex A to paper by Michael Ashmore, 18 Oct. 1993.

81. *The Times*, 14 and 25 Oct. 1995; Viscount Slim to the editor, *The Times*, 5 Jan. 1996; *Herald*, 1 Jan. 1996.

Notes to Chapter 10

1. Ruddock F. Mackay, *Fisher of Kilverstone* (Oxford, 1973), 289.

2. Fisher to Esher, *c.*7 Nov. 1903, in Arthur J. Marder (ed.), *Fear God and Dread Nought: The Correspondence of Admiral of the Fleet Lord Fisher of Kilverstone* (3 vols., London, 1952–9), i. 290.

3. David French, *The British Way in Warfare 1685–2000* (London, 1990), 59, 120.

4. Fisher to J. A. Spender, 27 Feb. 1911, in Marder, *Fear God and Dread Nought*, ii. 359.

5. John Gooch, 'The Bolt from the Blue', in *The Prospect of War*.

6. Steven R. B. Smith, 'Public Opinion, the Navy and the City of London: The Drive for British Naval Expansion in the Late Nineteenth Century', *War & Society*, 9 (1991), 29–50.

7. A. J. A. Morris, *The Scaremongers: The Advocacy of War and Rearmament 1896–1914* (London, 1984).

8. Major General Sir George Aston, *Memories of a Marine: An Autobiography* (London, 1919), 232, 237.

9. James Leasor, *War at the Top: Based on the Experiences of General Sir Leslie Hollis K.C.B., K.B.E.* (London, 1959), 34–5.

10. Marshal of the RAF Sir John Slessor, *The Central Blue: Recollections and Reflections* (London, 1956), 59–71.

11. Malcolm Smith, *British Air Strategy between the Wars* (Oxford, 1984), 181–2, 319–20; see also Andrew Boyle, *Trenchard* (London, 1962).

12. Bond, *Chief of Staff*, i. 21.

13. Quoted by Dr Maurice Pearton, in an unpublished paper on civil–military relations.

14. *Memoirs of Lord Ismay*, 165–6.

15. Kennedy, *The Business of War*, 76, 97.

16. Bond, *Chief of Staff*, ii. 19–22.

17. Brian Loring Villa, *Unauthorized Action: Mountbatten and the Dieppe Raid* (Toronto, 1989).

18. Kennedy, *The Business of War*, 206–7; also 231–2.

19. Slessor, *Central Blue*, 418–26; see also Bryant, *Turn of the Tide*, 595.

20. Franklyn Arthur Johnson, *Defence by Ministry: The British Ministry of Defence 1944-1974* (London, 1980), 18-21.

21. Hamilton, *Monty: The Field Marshal*, 687.

22. Field Marshal the Viscount Mongomery of Alamein, 'Organization for War in Modern Times', *Journal of the Royal United Service Institution*, 100 (1955), 509-31.

23. Leasor, *War at the Top*, 12-13; Hamilton, *Monty: The Field Marshal*, 642-4, 668, 677.

24. Ibid. 652-8, 688.

25. Montgomery to Sir John Anderson, 1947, quoted in *The Times*, 28 Mar. 1995; more generally, L. V. Scott, *Conscription and the Attlee Government: The Politics and Policy of National Service 1945-1951* (Oxford, 1993), esp. 16-19, 76-8, 92, 96; Ronald Lewin, *Slim: the Standardbearer* (London, 1976), 268-70.

26. Paul Nicholas Cornish, 'A Token Commitment: British Military Planning for the Defence of Germany 1945-50' (Cambridge University Ph.D. thesis, 1993), esp. 21-7, 170, 188, 201, 235, 295-301, 323-30, 383-93, 414-17; also Scott, *Conscription and the Attlee Government*, 172-5; Simon James Ball, 'The Royal Air Force and British Nuclear Strategy, 1945-59' (Cambridge University Ph.D. thesis, 1991).

27. Ian Clark and Nicholas J. Wheeler, *The British Origins of Nuclear Strategy 1945-1955* (Oxford, 1989), 160-74.

28. John Baylis, *Ambiguity and Deterrence: British Nuclear Strategy 1945-1964* (Oxford, 1995), 166-74, 193-6, 210-12.

29. William P. Snyder, *The Politics of British Defense Policy, 1945-1962* (Ohio, 1964), 130.

30. Simon J. Ball, 'Harold Macmillan and the Politics of Defence: The Market for Strategic Ideas during the Sandys Era Revisited', *Twentieth Century British History*, 6 (1995), 86.

31. Martin S. Navias, *Nuclear Weapons and British Strategic Planning 1955-1958* (Oxford, 1991), esp. 86-98, 141-57.

32. Abrams, 'Democracy, Technology and the Retired British Officer', 161; Andrew J. Pierre, *Nuclear Politics: The British Experience with an Independent Strategic Force 1939-1970* (London, 1972), 193-4; Drusilla E. Wiffen, 'Service Politics and the Sandys 1957 Defence White Paper', (thesis for M.Phil. in International Relations, Cambridge University, 1988), 61-2.

33. Public Record Office, DEFE 32/5, 7 Nov. 1957, quoted in Wiffen, 'Service Politics', 24.

34. Philip Ziegler, *Mountbatten: The Official Biography* (London, 1985), 528; in later versions the recipient of this remark became Duncan Sandys—see Denis Healey, *The Time of my Life* (London, 1989), 257.

35. Ball, 'Harold Macmillian and the Politics of Defence', 86-8.

36. The literature on the central organization of defence includes Johnson, *Defence by Ministry*; Michael Howard, *The Central Organisation for Defence* (London, 1970); Martin Edmonds, 'Central Organizations of Defence in Great Britain', in Martin Edmonds (ed.), *Central Organizations of Defense*

(Boulder, Col., 1985); Martin Edmonds, 'The Higher Organisation of Defence in Britain, 1945–85: The Federal-Unification Debate', in Edmonds, *The Defence Equation*; John Gooch, 'The Chiefs of Staff and the Higher Organization for Defence in Britain, 1904–1984', *US Naval War College Review* (Jan.–Feb. 1986), 53–65.

37. Cloake, *Tiger of Malaya*, 337–9, 359, 369–70.
38. General Sir Charles Richardson, *From Churchill's Secret Circle to the B.B.C.: The Biography of Lieutenant-General Sir Ian Jacob* (London, 1991), 272.
39. Ziegler, *Mountbatten*, 581–2, 608–9.
40. Ibid. 636.
41. Healey, *The Time of my Life*, 261.
42. Ibid. 263.
43. Johnson, *Defence by Ministry*, 186–7; Ewen Broadbent, *The Military and Government from Macmillan to Heseltine* (Basingstoke, 1988), 225.
44. Carver, *Out of Step*, 347, 398.
45. *The Times*, 4 Apr. 1981.
46. Hill-Norton to the editor, *The Times*, 13 May 1980; see also Carver, *Out of Step*, 543.
47. *The Times*, 10 Nov. 1983.
48. *The Times*, 6 Feb. 1985; Lewin to the editor, ibid. 25 Nov. 1991 and 18 Feb. 1992.
49. See, for example, Marshal of the RAF Sir William Dickson to the editor, *The Times*, 31 Mar. 1980; Marshal of the RAF Sir Neil Cameron to the editor, ibid. 9 and 29 May 1980.
50. Carver, *Out of Step*, 543–4.
51. Field Marshal Lord Carver to the editor of *The Times*, 26 June 1981; also 16 May 1980; Carver, *A Policy for Peace* (London, 1982); see also Carver's speeches in the House of Lords on 18 Dec. 1979 and 23 Apr. 1980, and at the Royal United Services Institute for Defence Studies, 30 Apr. 1980; House of Commons papers 1978–9, no. 348, *6th Report from the Expenditure Committee: The Future of the United Kingdom's Nuclear Weapons Policy*, evidence p. 211.
52. Bagnall to conference on sea power, Hull University, 30 Mar. 1994, and letter to editor, *The Times*, 3 Aug. 1994; Beach to editor of *The Times*, 9 July 1993.
53. General Sir William Jackson and Field Marshal Lord Bramall, *The Chiefs: The Story of the United Kingdom Chiefs of Staff* (London, 1992), 382; *The Times*, 11 Dec. 1976.
54. *The Times*, 31 Oct. 1978.
55. Ibid. 22 May 1981.
56. Hill-Norton to the editor, *The Times*, 19 Jan. 1981.
57. *The Times*, 22 May 1981.
58. Keith Speed, *Sea Change: The Battle for the Falklands and the Future of Britain's Navy* (Bath, 1982), 9–13.

59. Max Hastings and Simon Jenkins, *The Battle for the Falklands* (London, 1983), 61–70.

60. Max Hastings, *Spectator*, 23 Feb. 1985; Captain Nick Kerr, 'The Falklands campaign', *US Naval War College Review* (Nov.–Dec. 1982), 14–21.

61. John Nott, 'After the Falklands, Let's not Go Overboard on Navy Spending', *The Times*, 27 July 1982.

62. Carver to the editor, *The Times*, 23 June 1982; see also 28 June, 6, 8, 12 July.

63. *The Times*, 6 Oct. 1982.

64. John Nott, 'Our Defences All at Sea', ibid. 5 Oct. 1987.

65. Broadbent, *The Military and Government*, 62–3.

66. Ibid. 206–7.

67. David K. Boren, 'Establishing Civilian Supremacy: influence within Britain's Ministry of Defence', in Paul Smith (ed.), *Government and the Armed Forces in Britain 1856–1990* (London, 1996).

68. Lt.-Gen. Sir Maurice Johnston, 'More Power to the Centre', lecture to the Royal United Services Institute for Defence Studies, 12 Oct. 1982; Lewin to editor, *The Times*, 20 Mar. 1984.

69. *The Times*, 13 Mar. 1984.

70. Sir Edwin Bramall, 'The Contributions of the Chiefs of Staff to U.K. Defence', *Journal of the Royal United Services Institute for Defence Studies*, 131 (Sept. 1986), 9; see also Sir Clive Whitmore, 'Ministry of Defence Reorganisation: The Implementation of Change', ibid. 130 (Mar. 1985), 7–12; Jackson and Bramall, *The Chiefs*, 428–32.

71. House of Lords debate reported in *The Times*, 14 June 1984; Cameron to the editor, *The Times*, 17 May 1984; Hill-Norton, ibid. 30 Mar. 1984; Carver, ibid. 23 Mar. 1984; Rodney Cowton, 'Can Brass Deflect a Heseltine Misguided Missile?', ibid. 14 May 1984.

72. *The Times*, 21, 23 May 1984.

73. Ibid. 9, 19, 20, 26 July 1994.

74. Jackson and Bramall, *The Chiefs*, 450.

75. Broadbent, *The Military and Government*, 107.

76. Michael Hobkirk, 'The Heseltine Reorganisation of Defence: Kill or Cure?', *Journal of the Royal United Services Institute for Defence Studies*, 130 (Mar. 1985), 45–50; Hobkirk, 'Reform across the Sea: A Comparison of Defence Policy Making in the U.K. and the U.S.A.', ibid. 132 (Sept 1987), 55–60.

77. *The Economist*, 19 May 1990, p. 35; *The Times*, 18, 20, 28 June 1990, 10 July 1991; David Bolton, 'Defence in Transition: Options for Change', *Journal of the Royal United Services Institute for Defence Studies*, 136 (Autumn 1991), 1–3. For a somewhat different and more egocentric account of the review's genesis, see Alan Clark, *Diaries* (London, 1993), 257–66, 287–8, 298–300, 304, 309.

78. *The Times*, 15 July 1994.

79. Ibid. 9 and 10 Nov. 1993.

80. Frank Kitson, 'Smaller Army with a Bigger Punch', *The Times*, 31 July

1990; Jackson to the editor, *The Times*, 2 Aug. 1990; Bramall, ibid. 7 Aug. 1990.

81. Field Marshal Sir Nigel Bagnall, 'Keeping up our Guard against the Unexpected', *Daily Telegraph*, 26 June 1990.

Notes to Conclusion

1. Johnson, *Defence by Ministry*, 181; for similar surveys see Lt.-Col. J. C. M. Baynes, *The Soldier in Modern Society* (London, 1972), 56–65; David Capitanchik, 'Public Opinion and Popular Attitudes towards Defence', in John Baylis (ed.), *British Defence Policy in a Changing World* (London, 1977).

2. David Capitanchik and Richard C. Eichenberg, 'Defence and Public Opinion', *Chatham House Papers*, 20 (1983), 19.

3. Lance E. Davis and Robert A. Huttenback, *Mammon and the Pursuit of Empire: The Political Economy of British Imperialism 1860–1912* (Cambridge, 1986), 160.

4. Alfred Vagts, *A History of Militarism: Civilian and Military* (London, 1959); see also Volker R. Berghahn, *Militarism: The History of an International Debate 1861–1979* (Cambridge, 1984); John R. Gillis (ed.), *The Militarization of the Western World* (New Brunswick, 1989).

5. David Edgerton, 'Liberal Militarism and the British State', *New Left Review*, 185 (Jan.–Feb. 1991), 138–69; see also David Edgerton, *England and the Aeroplane: An Essay on a Militant and Technological Nation* (London, 1991).

6. Johnson, *Defence by Ministry*, 150.

7. Sir Arthur Hockaday, 'Parliamentary Control of Defence: The Role of Select Committees', *Journal of the Royal United Services Institute for Defence Studies*, 135 (Spring 1990), 7–10.

8. Mackintosh to the editor of *The Times*, 8 Jan. 1977; see also John P. Mackintosh, 'Britain's Defence Cuts may be Putting the west Seriously at Risk', ibid. 4 Apr. 1977.

9. For evidence of one minister's frustrations, see Clark, *Diaries*, 290–1, 306, 395.

10. Baynes, *No Reward but Honour?*, 62; see also Denaro, 'Warrior or Worker', 42.

11. Huntington, *The Soldier and the State*, 98–124; John W. Wheeler-Bennett, *The Nemesis of Power: The German Army in Politics 1918–1945* (London, 1953).

12. Sir Peter de la Billière, *Looking for Trouble: An Autobiography: From the S.A.S. to the Gulf* (London, 1994), as quoted by Alistair Horne in *The Times*, 20 Oct. 1994.

13. *Scotsman*, 16 Jan. 1992.

14. *The Times*, 6 Oct. 1986, and 9 June 1987.

15. Ibid. 14 Jan. 1990.

INDEX